Renewable Power
in Developing Countries
Winning the War on Global Warming

Renewable Power in Developing Countries
Winning the War on Global Warming

Steven Ferrey

with

Anil Cabraal

HD
9502
.D44
F47
2006
West

PennWell

DISCLAIMER

Information has been obtained from sources that are believed to be reliable. Neither the author nor any of the quoted sources make any warranty, express or implied, or assume any legal liability or responsibility for the accuracy, completeness or usefulness of any information, apparatus, product or process disclosed in this book, or represent that the use of any such information, apparatus, product or process would not infringe privately owned rights.

Copyright ©2006 by PennWell Corporation
1421 South Sheridan Road
Tulsa, OK 74112-6600 USA
800-752-9764
www.pennwellbooks.com
www.pennwell.com

Director: Mary McGee
Managing Editor: Steve Hill
Production/Operations Manager: Traci Huntsman
Production Manager: Robin Remaley
Assistant Editor: Amethyst Hensley
Production Assistant: Amanda Seiders
Book Designer: Clark Bell

Library of Congress Cataloging-in-Publication Data Available on Request

Ferrey, Steven with Anil Cabraal
Renewable Power in Developing Countries: Winning the War on Global Warming

ISBN 1-59370-050-4

All rights reserved. No part of this book may be reproduced, stored in a retrieval system, or transcribed in any form or by any means, electronic or mechanical, including photocopying and recording, without the prior written permission of the publisher.

Printed in the United States of America

1 2 3 4 5   10 09 08 07 06

Contents

Acknowledgments ix

Introduction xi
   What Matters xi
   What Follows xiv

Exchange Rates xix

Acronyms, Abbreviations, and Definitions xx

About the Authors xxv

1 Critical Development in Asia 1
   The Best of Times, the Worst of Times 1
   Infrastructure Lifetimes and Hard Choices 22

2 Renewable Energy as the Global Warming Solution 35
   Renewable Energy Options and Alternatives 35
   The Renewable Technologies for Developing Nations 37
   Comparative Environmental Emissions 46
   Decentralization of Electric Generation 54
   Distributed Generation Reliability 55

3 Overview: What Works for Renewable Power
   Implementation in Developing Nations 67
   Thailand Program Summary 74
   Indonesia Program Summary 76
   India Program Summary 78
   Sri Lanka Program Summary 80
   Vietnam Program Summary 82

4 Thailand: Creative Competitive Markets
   in the Heart of the Tiger 85
   Program Overview 85
   Program Design and Implementation 86
   Power Purchase Agreements 91

5 Indonesia: Carrots Rather than Clubs—
Incentives for Peak Performance 99
Program Overview 99
Program Design and Implementation... 100
Power Purchase Agreements 107

6 India: State Power in a Federalist System.................................. 119
Program Overview 119
Andhra Pradesh 125
Tamil Nadu...134

7 Sri Lanka: 21st-Century Conversion of
Ancient Renewable Technologies 145
Program Overview 145
Program Design and Implementation..146

8 Vietnam: Capital Markets and Renewable Energy
in a People's Republic 159
Program Overview 159
Supply Resources and Demand..159
Institutional Differences in the Vietnamese System 161
Program Design and Implementation..164
Power Purchase Agreements 175

9 Lessons Learned in Asia for a Successful SPP Program:
Template and Techniques 185
Common Threads and Differences 185
Key Issues in Renewable SPP Program Design...................................187
Recommended Best Practices and Program Template....................194

10 Financing the Transition 201
Transitions to Restructured Markets... 201
The Magnitude of Required Electric-Sector Investment...................202
Funding Sources...205
Electric-Sector Investment Criteria 206
Phases of Financial Risk 208
Types of International Project Risk 210
Shifting and Bearing Risk in the Power Sector 213
Risk Mitigation.. 215
Government Role and Reform for
    Small Power Producer Programs.. 218
Reform of the Existing Legal Framework 221

	International Credit Agencies and Their Financing Functions	223
	Other International Credit Agencies and Sources	231
	Risks of Renewable Energy Development	233
11	**Key Provisions in Power Purchase Agreements**	**247**
	Overview	247
	Contract Formation and Contract Validity	251
	Modification of the Contract	257
	Dispute Resolution	261
	Breach of the Contract	262
	Liability	269
	Lenders' Rights	270
12	**The Law and Principles Governing International Power**	**275**
	Securing Power Contract Enforcement	275
	Form, Formation, and Modification of Power Sale Contracts	281
	Interpretation of the Deal	288
	The Obligation to Honor the Power Purchase Agreement	292
	Conclusion	296
13	**The Eye of the Needle: International Environmental Assessments and Reviews for Power Projects**	**311**
	Types of Projects Covered	314
	Project Screening	315
	Responsibility for Environmental Assessment Preparation	317
	Initiation of the Environmental Assessment Process	319
	Project Scoping	320
	Type of Impacts Considered in the Review	322
	Consideration of Project Alternatives	325
	Mitigation Discussion or Adoption of Mitigation	326
	Timing of EIA Document Preparation	327
	Environmental Assessment Review and Public Participation	329
	Final Decisions to Proceed with Energy Projects	330
	Postapproval Monitoring and Auditing	331
	Conclusion	333
Index		**343**

Acknowledgments

This book is the culmination of more than a decade of our work in developing nations in Asia, Africa, and Latin America. Steven Ferrey and Anil Cabraal collaborated on chapters 3 through 9 documenting the specific experience of the Asian nations. Professor Ferrey is responsible for creating chapters 1 and 2 and 10 through 13.

The authors gratefully acknowledge the help of all those who assisted with the research on these issues, including P. R. K. Sobhan Babu, A. Reddy, and S. V. R. Rao at Winrock, India; Robert Vernstrom, Pom Vattasingh, and their associates in Thailand; Romesh Dias Bandaranaike in Sri Lanka; Pham Nguyet Ahn at the World Bank office in Hanoi, Vietnam; as well as all those ministry officials, utility personnel, project developers, lenders, and others in each of the five Asian nations studied herein who generously provided information and gave their time for interviews with Professor Ferrey. Professor Ferrey thanks the Asia Alternative Energy Program; the World Bank, for supporting some of the study that became the inspiration for this book; as well as Matthew Mendis, Ann Herendeen, and the staff at International Resources Group, Ltd., who have allowed material from studies to be part of this book. The author thanks Dean Robert Smith, and Associate Deans Marc Perlin and Bernard Keenan at Suffolk University Law School for their kind and generous support and encouragement during the summer of 2004 for the writing of this book. Professor Ferrey also thanks his research assistants, who assisted in the completion of this book: Douglas Martland, Ari Bessendorf, and Eric Amundson. Despite all of the assistance received, any errors or omissions in this evaluation are solely the responsibility of Professor Ferrey.

Introduction

What Matters

If the crisis is global, then the solution must include developing nations, particularly those in Asia. This book charts a course that may be the only way to advance—and ultimately to win—the battle against global warming.

Global warming confronts world leaders with the environmental challenge of this century. According to many scientists, every additional ton of carbon dioxide, as well as five other *greenhouse gases* (GHGs), emitted by the burning of fossil fuels (coal, oil, gas) is warming the planet to potentially dangerous and irreversible levels. The primary culprit is the burning of copious quantities of fossil fuels for the production of electric power around the world. No matter where it is expelled, a ton of released carbon dioxide from a conventional electric power plant imposes a similar threat to all citizens of the globe. If the climate warms, the effects are universal.

Conventional electric power production is the focal point of reform. The demand for more electric power is almost insatiable, especially in developing nations that are just now realizing the consumer and investment benefits of this technology. There is no turning back the strong demand of people in developing nations to have the benefits of progress that electricity fosters: On a personal level, electricity has been essential in easing the lives of the traditionally disadvantaged half of the humanity as it did away with tiresome domestic labor and offered the possibility of female emancipation."[1]

Electricity has no peers for lighting (which illuminates nighttime activities, including allowing children to study during the evening hours) and for such basic services as elevators and air conditioning. Electric power is essential for telephone networks, the Internet, and wireless communications. For equatorial countries, which contain the bulk of the world's population, the ability to electrically air condition space is a prerequisite for commercial development. Lee Kuan Yew, founder and former president of Singapore, identified the air conditioner as the most important invention of the 20th century for the country's development, which is often regarded as a prototypical success story for developing nations.[2]

The responsibility for addressing global warming must be shared by all nations and people as a common but differentiated responsibility. The developed nations produce the bulk of GHGs today; developing nations are the burgeoning source of the future. Despite the tack of the Kyoto Protocol to impose the entire GHG reduction obligation on developed nations, this alone will not solve the problem. The battle against global warming can be won only by including developing nations, especially those in Asia, in the solution. Why is this so when the developed countries have the much higher per capita energy consumption and GHG emissions?

The future is denominated by significant changes at the margin of our energy base. Energy use is a function of population, development, and technology choice. Population and development are key increasing variables in many developing countries. Technology is a choice.

Population

Population is a key determinant of electric energy use. World population could reach eight billion people by 2020 and 9–10 billion by 2050. Electricity demand growth is increasing at a significant rate in Asia. Currently, Asia is home to three of the world's four most populous countries. Of the current six billion in world population, more than half live in Asia.[3]

Many of these people do not yet contribute to modern energy consumption. This poorest one-quarter of humanity now uses less than 5% of all commercial primary energy, but this proportion will increase.[4] There are almost two billion people in the world, primarily in South Asia and sub-Saharan Africa, with no access to electricity, but who seek it.[5] Many others in poor countries with access to electricity cannot yet afford to purchase it, but in the future they will.

Development

Sixty percent of all future GHGs will be emitted in Asia, more than all other continents combined. Between 1980 and 2002, China's installed electric generation capacity grew from 65 GW to 353 GW, making it the second-largest generating base in the world. The global environmental challenge of the 21st century will be won or lost from the ground up—and it must be won in developing nations, particularly in Asia, to be won on a global basis.

Developing Nations

Developing nations, particularly in Asia, must be in the mix for any solution. The logic is simple: Although the United States attracts international attention because of its consumption of more than 25% of global fossil fuels and a commensurate emission of GHGs, to focus only on developed nations is like watching ourselves in the rearview mirror when the future lies ahead. Even if the United States were to achieve the targets set out in the Kyoto Protocol—to reduce GHG emissions to 7%–12% below 1990 levels by 2012—that would mean a reduction of about 18% of the current U.S. share, a reduction of less than 5% of global GHG emissions. Although this would be an important achievement, it is notable that GHG emissions have increased progressively rather than decreased during the Bush, Clinton, and second Bush administrations.

Not only will more than 60% of all future increases in GHG emissions emanate from the single continent of Asia, but within 15 years of the publication of this book, China alone is expected to double its GHG emissions and quadruple its GHG output before 2050. This would swamp the reductions that the United States might make if it achieves the Kyoto targets. And this is just China. Add the burgeoning GHG emissions from India, which, with its prodigious growth rate, is on the way to becoming the most populous nation on Earth; Indonesia, the fourth most populous nation in the world, with its high birth rate; and a multitude of fast-developing nations such as Thailand and Vietnam, each of which already exceed the populations of major European nations such as France, Britain, and Germany. A key battlefront on global warming is fixed in Asia.

Solutions

Amid the often pessimistic news on global warming, there is good news, as well as viable solutions. Alternative renewable energy technologies are available and feasible today for electric energy production. Renewable energy can replace and displace fossil fuel use. Moreover, there are proven and viable models for renewable energy development from the ground up. These include electric-sector reform in developing nations. This book is a roadmap to what has been tried and what has and has not worked in implementing renewable energy alternatives in critical developing nations.

Moreover, rapid electric growth in developing nations is not only an alarm, but an opportunity. With a need for substantial new power resources in developing nations, one can decide what types to deploy and where to site them. With many rural areas not yet served with electric power, there is a choice as to whether to run transmission lines to these areas, or instead put these funds toward the higher initial costs of stand-alone renewable energy systems to meet future power needs. In these choices, the future fate of the planet's temperature may lurk.

What Follows

This book charts the renewable energy successes and failures in five key developing nations in Asia. It creates a blueprint of recommendations for mitigating global warming from the ground up. We sketch the problem, the challenge, and the solution.

Chapter 1 explores the current science and policy issues of global warming from both sides of this often polarized debate. It charts the differences between renewable and conventional electric generation technologies in their emission of greenhouse gases that affect global climate. The chapter concludes by charting a discernible path for implementing immediate policy choices in developing nations.

The renewable and distributed technology options for implementing this path are set forth in chapter 2. It continues to explore the environmental impacts beyond just greenhouse gases when deploying renewable electric energy technologies in lieu of traditional fossil fuel combustion.

From an understanding of the GHG challenge and technology alternatives, one needs to identify practical, reasonably implemented policy solutions. The next chapters analyze the renewable power programs in key developing Asian nations. Chapter 3 summarizes what has and has not worked for sustainable renewable energy projects in Asia. Subsequently, we analyze in detail the programmatic, legal, regulatory, and tariff aspects of the Asian experience with sustainable energy policies and programs and the lessons they offer for developing nations.

Thailand (chapter 4) features the original Asian independent renewable power program. Thailand represents a relatively mature developing nation with a population size (65 million) greater than the populations of England, France, or Germany. It features a relatively mature system of law. Most important for our analysis, Thailand illustrates how competition can be introduced into sustainable energy programs, even when there is a development path that does not rely on direct international subsidies.

Indonesia (chapter 5) is the world's largest Muslim nation with the luxury of significant oil and gas resources. Indonesia offers the particular challenge of a nation composed of thousands of inhabited islands, such that each area must be served by its own freestanding power grid. This disaggregation of the electric infrastructure opens up creative program opportunities for sustainable energy options. One innovative element of the Indonesian system is a disaggregated tariff and a competitive solicitation for projects on different islands that reflects regional variations. Moreover, the Indonesian scheme employs steeply tilted tariff incentives in lieu of legal compulsions that otherwise would be embodied in the power purchase agreement.

India (chapter 6) is the world's largest democracy and a key example of a developing nation implementing its electric energy sector at the state level through a federalist system, as in the United States. India, unlike some developing nations, has an established system of law and has used its programs as a fulcrum to become a regional manufacturing center for renewable energy generating equipment. In India, we compare and contrast how two neighboring Indian states have implemented a renewable energy program. What is innovative is a tariff structure that recognizes the contribution of sustainable energy.

Sri Lanka (chapter 7) is an example of a smaller developing nation without any significant indigenous sources of fossil fuels. It has conducted one of Asia's most successful renewable energy programs amid a 20-year civil war. Distributed and renewable resources may prove its most resilient and reliable electric energy sources. Sri Lanka offers the interesting opportunity to convert historical 19th-century tea plantation hydro resources, which were originally employed for mechanical power, to hydroelectric energy production for the 21st century. Sri Lanka demonstrates the use of a noncompetitive open offer for renewable power resources rather than a competitive-bidding process.

Vietnam (chapter 8) opens a window on renewable energy in a developing communist/socialist nation. A renewable energy program in such a system, as in Vietnam or China, is unique in that the developers are state-owned or state-controlled companies, capital flows are state directed, and the role played by contracts, tariffs, and institutional program are correspondingly distinct. Chapter 8 examines the pending reforms to these systems to implement sustainable technologies as a meaningful part of the electric generation mix.

Chapter 9 draws together all of the lessons learned from the comparison of these key Asian developing nations' experience with small power producer (SPP) renewable energy development. It summarizes what has worked, why certain initiatives have failed, and what are best management program elements and enhancements applicable to all developing nations. It creates a template for developing nations based on what has and has not worked for renewable energy. It concludes with recommended best techniques for any developing nation.

Setting up the program properly in-country is a necessary, but not a sufficient element for renewable SPP program success. International financial market considerations, international environmental requirements, a properly structured power purchase agreement (PPA) between the utility

Introduction

buyer and the SPP power seller, as well as the often invisible legal elements for enforcement of the PPA, must all align seamlessly for international capital to flow to the electric sector. The final chapters of the book examine each of these necessary elements. The final four chapters analyze the requisites and mechanics of these international elements.

Chapter 10 analyzes the mechanics of international finance and risk allocation in renewable energy development. Renewable energy development requires that the requisites of international finance be carefully observed. The role of international agencies and banks is examined; the benefits of different international credit, grant, and lending opportunities are compared strategically.

Energy projects are more than steel and cement. Energy projects begin and end with the intangible assets of the deal. The power deal is embodied in the PPA. Design it incorrectly, and the project will never be financed or completed. Chapter 11 analyzes the key provisions of power purchase agreements that become the living legal skin of international renewable energy projects. We offer an analysis of the requisite and proper PPA provisions for agreements with independent power producers and SPPs.

Disputes occur in international power deals. Even a properly structured PPA does not resolve all of the key legal issues. The choice substantive law to govern the transaction and resolve any disputes is a crucial determination. It is not always as transparent as it may seem. The choice of law varies not only among nations, but may vary within a national legal system, depending on which branch of the law applies. Chapter 12 compares the international legal provisions, choice of law, and decision rules applicable to dispute resolution in renewable energy development.

Environmental issues stop otherwise well conceived energy projects. At a minimum, environmental requirements always pose a significant obstacle course to be navigated by any project. Chapter 13 sets forth the international environmental criteria for electric energy projects in developing nations. Projects must satisfy international environmental criteria to be eligible for international capital or aid. We contrast the different environmental evaluation standards that apply to various sources of international funding and credit guarantees.

Assembling all of these pieces creates a complete picture of successful SPP projects in developing nations. In addition to viable program design, fair and neutral PPAs and avoided cost-based tariffs, financing, and environmental considerations are necessary components of successful project execution. Successful international SPP programs incorporate a set of complementary and increasingly standardized provisions. These are necessary to facilitate international capital flows to the power sector.

An international form for power financing is emerging. In this form is function; the antidote to global warming and models for renewable electric implementation in developing countries will be forged in the success of renewable and sustainable energy development in developing countries, particularly in Asia, where a majority of new GHGs will otherwise be emitted. From these experiences are the transcendent lessons for our energy future.

We start in chapters 1 and 2 with an examination of global warming and the significant differences between the environmental impacts of renewable and traditional fossil fuel technologies for the generation of electric power. The unavoidable reality we confront is that demand for more power generation is exploding in developing countries around the globe. With more conventional electric power use in both developed and developing countries, the global warming has become an international problem. As a world community, we must address it deliberately and intelligently now, or else atomistically remain adrift amid the swelling greenhouse gas currents. To ourselves, and to future generations, we owe only the best, most intelligent, and sustainable power development.

Notes

1. Smil, Vaclav. 1999. *Energies: An illustrated guide to the biosphere and civilization*, Cambridge: MIT Press.

2. Wayne Arnold. 2002. Singapore cools off, and all must pitch in. *New York Times*, June 2, 6. "Air-conditioning plays a crucial role in our economy. Without it many of our rank-and-file workers would probably be sitting under coconut trees to escape from the heat and humidity instead of working in high-tech factories," stated Lim Swee Say, Singapore's environmental minister. In Singapore, air conditioning accounts for nearly 10% of the cost of new buildings and approximately 60% of monthly power bills.

3. 1.8 billion people live in East Asia and China, and 1.3 billion live in South Asia.

4. Smil, *Energies*, 197.

5. International Energy Agency (IEA). 2002. *World energy outlook 2002*. Paris: IEA.

Exchange Rates

	(as of June 2003)	(as of October 2005)
1 U.S. dollar =	42.77 Thai baht	41.217 Thai baht
1 U.S. dollar =	47.52 Indian rupees	43.95 Indian rupees
	1 lakh = 100,000 rupees	
	1 crore = 10 million rupees	
1 U.S. dollar =	96.86 Sri Lankan rupees	101.52 Sri Lankan rupees
1 U.S. dollar =	8,845 Indonesian rupiah	10,252 Indonesian rupiah
1 U.S. dollar =	15,426 Vietnamese dong	15,888 Vietnamese dong

Note: Both 2003 and 2005 currency exchange rates are set forth; June 2003 exchange rates are utilized for currency conversions in the text.

Acronyms, Abbreviations, and Definitions

ADB	Asian Development Bank
AfDB	African Development Bank
APERC	Andhra Pradesh Electricity Regulatory Commission
APSEB	Andhra Pradesh State Electricity Board
APTransco	Transmission Corporation of Andhra Pradesh Limited
ASTAE	Asia Alternative Energy Program
Avoided cost	The cost that a purchaser would pay if it generated or purchased power from external sources or at market rates
BOOT	Build–own–operate–transfer
Capacity	Electric generation capacity firmly committed for a period of time
CEB	Ceylon Electricity Board (Sri Lanka)
CH_4	Methane
CISG	Convention on the International Sale of Goods
CO	Carbon monoxide
CO_2	Carbon dioxide
CO_{2e}	Carbon dioxide equivalents
Cogeneration	The production of electricity and heat from a common source
DANIDA	Danish International Development Agency
DGEED	Directory General of Electricity and Energy Development (Indonesia)
Dispatchable power	Power from an electric power generator that can be controlled, usually by the purchasing utility as to its time or quantity of power output
EAP	Environmental Action Plan
EA	Environmental assessment
EMP	Environmental Management Plan
ENCON	Energy Conservation Promotion Fund Committee (Thailand)

Energy	Kilowatt-hours of power not firmly committed long term to a particular use or sale
EGAT	Electricity Generation Authority of Thailand
EPO	Energy Policy Office (Thailand)
ESRP	Environmental and Social Review Procedure (IFC, MIGA)
EVN	Electricity of Vietnam
EIS	Environmental Impact Statement
EBRD	European Bank for Reconstruction and Development
ESMAP	Energy Sector Management Assistance Programme, a part of the World Bank
FDR	Fixed deposit receipt
Firm power	Power that is contractually committed to a particular use or purpose
FS	Feasibility Study
GDP	Gross domestic product
GEF	Global Environment Facility
GHG	Greenhouse gases, including carbon dioxide
Gigawatt (gW)	1 million kilowatts (kW)
Gj	Gigajoules, 1 million joules
GNP	Gross national product
GWh	Gigawatt-hour
HDI	Human Development Index
HFCs	Hydrofluorocarbons
IBRD	International Bank for Reconstruction and Development (World Bank)
ICSID	International Centre for Settlement of Investment Disputes
IDA	International Development Agency
IEA	International Energy Agency
IEC	Electro-technical Commission
IFC	International Finance Corporation
IFI	International financial institution
IPCC	Intergovernmental Panel on Climate Change
IPP	Independent Power Producer

IREDA	Indian Renewable Energy Development Agency
IUKU	Electric Power Enterprise Permit (Indonesia)
KfW	Kreditanstalt für Wiederaufbau (Bank for Reconstruction; Germany)
Km	Kilometers
kW	Kilowatt
kWh	Kilowatt-hour
LoI	Letter of Intent
Marginal cost	The cost of providing the next kilowatt-hour or kilowatt of power, or both
MNES	Ministry of Non-Conventional Energy Sources (India)
MW	Megawatt
MWh	Megawatt-hour
MDB	Multilateral Development Bank
MFI	Multilateral financial institution
MIGA	Multilateral Investment Guarantee Agency
NASA	National Aeronautics and Space Administration (U.S.)
NEDCAP	Non-Conventional Energy Development Corporation of Andhra Pradesh (India)
NEPA	The National Environmental Policy Act (U.S.)
NGO	Nongovernmental Organization
NMOG	Nonmethane organic gases
Nonfirm power	Power that is not dedicated long term to a particular buyer and which the seller has no contractual obligation to produce
N_2O	Nitrous oxide
NO_x	Nitrogen oxides
O_3	Ozone
O&M	Operation and maintenance
OECF	Overseas Economic Cooperation Fund (Japan)
Off-peak	Periods of relatively low system demand by day or by season
OPIC	Overseas Private Investment Corporation
OTEC	Ocean thermal energy conversion

PCF	Prototype Carbon Fund
PCG	Partial credit guarantees
Peak	Periods of relatively high system demand by day or by season
PFCs	Perfluorocarbons
PLN	Pursahan Listrik Negara (Indonesian State Electricity Corporation Ltd., Indonesia)
PPA	Power purchase agreement
PRG	Partial risk guarantee
PRI	Political risk insurance
Prime mover	The technology to convert chemical energy to mechanical energy, such as gas turbines or combined cycle units
PURPA	Public Utility Regulatory Policies Act, a U.S. energy statute
PV	Photovoltaic(s)
QF	Qualifying facility, a legal designation for SPPs and cogenerators under the U.S. law PURPA
RE	Renewable energy
RESU	Regional Environmental Sector Unit (IBRD)
Renewables or renewable energy	Renewable energy sources that can be replenished by naturally occurring processes in daily cycles
Retail competition	A system in which more than one electrical provider can sell to retail customers and such customers have choice of their providers
Retail wheeling	The ability of a power generator to use the national transmission grid to move power to itself (self-service wheeling) or others at locations remote from the generator
Rp	Indonesia rupiah
Rs	Indian rupee
SF_6	Sulfur hexafluoride
SL Rs	Sri Lankan rupee
SO_2	Sulfur dioxide
SPP	Small power producer
T&D	Transmission and distribution
TNEB	Tamil Nadu Electricity Board (India)

TRANSCO	Transmission company
UCC	Uniform Commercial Code
UNDP	United Nations Development Programme
UNEP	United Nations Environment Programme
UNFCCC	United Nations Framework Convention on Climate Change
UNIDROIT	A code of international contract laws
VOCs	Volatile organic compounds
WBG	World Bank Group
Wheeling	The transmission of electricity by an entity that does not own the power

About the Authors

Steven Ferrey is a professor of law at Suffolk University Law School and an adjunct professor of law at the Boston University School of Law. He was a visiting professor of law at Harvard Law School during 2003. He teaches contract law, energy law, and environmental law. He consults internationally and in the United States on energy and environmental matters.

Professor Ferrey is the author of five books, including *The Law of Independent Power*, now in its 22nd edition, a three-volume book used throughout the world by energy law firms, governments, and regulators involved in the power sector (West, 2005). His books also include *The New Rules: A Guide to Electric Market Regulation* (PennWell, 2000), and *Environmental Law: Examples and Explanations* (3rd edition, Aspen, 2004), as well as more than 70 articles and book chapters about energy and environmental policy. He holds a bachelor's degree in economics from Pomona College and juris doctorate and master's degrees from the University of California at Berkeley. He was a postdoctoral Fulbright Fellow in energy studies at the University of London.

Professor Ferrey has served as a consultant for more than a decade on projects sponsored by the World Bank and the United Nations to establish the small power project programs and standardized power purchase agreements for Indonesia, Sri Lanka, Vietnam, and Uganda. He has served on advisory committees to the president of the United States, the U.S. Congress, the U.S. Department of Energy, and the U.S. Department of Environmental Protection, among other agencies. The author may be reached by e-mail in Boston at sferrey@suffolk.edu or by telephone at (617) 573-8103.

Anil Cabraal is a lead energy specialist and leader of the Renewable Energy Cluster in the Energy and Water Department of the World Bank in Washington DC. He was the renewable energy team leader at the Asia Alternative Energy Program at the World Bank, and is leading a bank-wide initiative, "Towards a Sustainable Energy Future: A renewable Energy Action Plan" that is expected to result in at least doubling the World Bank's financial commitment for renewable energy and energy efficiency within five years. During the past 10 years at the World Bank he was responsible for design and implementation of renewable energy projects in Indonesia, India, Sri Lanka, China, Philippines, Vietnam, and Papua New Guinea. He led the preparation of a series of training manuals, entitled *Quality Processes for Photovoltaics* which is now in its second edition. Dr. Cabraal also is the principal author of a World Bank technical paper, "Best Practices for Photovoltaic Household Electrification Programs," as well as numerous papers related to photovoltaic applications, renewable energy policy and market development in developing countries. He is a member of the executive boards of PV Global Approval Program, and the Triodos Renewable Energy for Development Fund. Anil Cabraal earned his doctorate in agricultural engineering from the University of Maryland, in 1978.

1 Critical Development in Asia

"Energy flows underlie all human activity and substantially influence both the economic and the ecological systems locally and regionally, as well as globally."[1]

The Best of Times, the Worst of Times

Global warming has been called a pending environmental Armageddon, the cause of catastrophic ecological devastation of the planet, and even an overblown tempest in a teapot. Although a significant segment—but certainly not all—of the scientific community has coalesced around the theory that global warming is the environmental challenge of our lifetimes, policymakers around the world articulate ambitious goals to reduce the threat of global warming on the ecosphere of the earth. Regardless of the accuracy of the science, which won't be fully proved until we are much further along, the policy focus to mitigate global warming is forming as the most significant environmental initiative of the 21st century.

To get started, the first part of this chapter reviews the science on both sides of the global warming debate. In terms of effective political action, it does not matter which branch of the scientific debate ultimately is correct. Because the true threat and ultimate impact of global warming will remain unknown until well into the 21st century when any warming repercussions are upon us, and it takes decades to alter the well-entrenched world patterns of combustion of fossil fuels in power plants and vehicles, the world must either make an immediate commitment to dampen greenhouse gas emissions, or continue business as usual. Under either scenario, future generations will reap the consequences that we now sow.

Under even the most optimistic scenarios to redirect the infrastructure of energy use to more sustainable technologies, if one does not initiate redirection in earnest, our global *boat* cannot be turned in time should global warming predictions prove true. By the time the global warming predictions are confirmed empirically, it will effectively be too late to reverse embedded technologies to avert consequences. How should policy makers now factor in scientific uncertainty? Under the *precautionary principle*, when long-term environmental impacts of anthropogenic sources of emissions are unknown, policy makers err in favor of environmental protection; one minimizes the probability of the maximum negative outcome. Because the earth's ecology is an extraordinarily complex living system not easily or cheaply—if at all—repaired after the fact, one might err to choose policies in favor of preserving the healthy balance of functioning natural systems.

The equation of global warming

As touched on in the introduction, the emission of greenhouse gases (GHG) is driven by a fairly simple relationship. Fig. 1–1 describes the formative relationship of technological choices and the degree of development in causing the increased emission of greenhouse gases.

$$\text{GHGs} = f\{[\text{Population}] \times [\text{degree of development/electrification}] \times [\text{choice of technology}]\}$$

Figure 1–1: Equation of global warming

Affecting any of the three elements in this equation changes the emission of atmospheric gases that drive the models of global warming. Which of these three variables can we influence by policy changes?

Population. Although population control has its proponents, there is little doubt that world population will increase significantly during the next 50 years, especially in less-industrialized, poorer, developing nations. World population could reach 8 billion people by 2020 and 9 to 10 billion by 2050.[2] East Asia (including China) and South Asia now contain more than 3 billion of the world's 6 billion population.[3]

There is a demand for energy services. Over the last 25 years, more than 1 billion people in developing nations have gained access to modern energy fuels, including electricity, coal, kerosene, natural gas, and liquefied natural gas.[4] There are, however, still between 1.6 billion to 2 billion people[5] worldwide without access to modern energy services or electricity; 56% of the world's rural population lacks access to energy services.[6] Because of population growth, the number of people without access to modern energy services is increasing at approximately 30 million people per year.[7] Under business-as-usual scenarios, the International Energy Agency (IEA) projects that in 25 years, even with $16 trillion in energy investments in developing nations, 1.4 billion people will still lack access to electricity.[8]

Today the world's poorest two billion of the six billion world population live on less than $1 per day and consume 0.2 tons of oil equivalent per capita; the one billion richest people use 25 times more.[9] Despite hurdles, low-income households are willing to spend a greater percentage of household income on energy services than higher-income households.[10] Poor households in developing countries spend between 10% and 20% of annual income on energy costs, whereas affluent households spend less than 2%.

There is a wide disparity between per capita energy (not only electric) consumption in developed countries compared to developing countries. For example, the World Energy Assessment indicates that North Americans consume 281 GJ per capita and Europeans consume 142 GJ per capita, whereas per capita consumption in Asia is only 28 GJ.[11] The richest 20% of the world population consume 58% of total energy resources, whereas the poorest 20% consume less than 4% of resources.[12]

Per capita electric consumption will increase dramatically with development in developing nations. To cope with the increased electrification that accompanies the substantial increase in per capita energy use, which will occur in developing nations in the next decades, the world may have to achieve a reduction of carbon dioxide (CO_2) of up to 50% during the 21st century.[13] Extending energy services to 10 million people per year requires approximately $10 billion dollars of annual sector investment.[14] This includes costs of providing new generation facilities, development of regional transmission systems, local distribution systems, and individual household connections. Transmission networks for electricity are costly to build and thus require sufficient population density to generate enough demand and revenues to cover the lost of line extension. Low-income rural areas may lack the requisite density, making them ineligible for on-grid service extensions. The capital outlay required for connecting to the distribution system is another major barrier to universal access to modern energy services. Low-income households are typically unable to purchase transmission network connections because it costs from $50 to $1000.[15]

Development/Electrification. These patterns of electric intensity are changing quickly in the 21st century. Forecasts suggest a very strong demand growth for electric energy over the next 50 years in developing countries. The average annual growth rate in developing countries' primary energy use from 1990 to 2001 grew by 3.2% per year, compared in industrialized countries to 1.5% over the same period.[16] The IEA predicts as much as an annual 4% increase in demand by developing countries over the next 20 years.[17] Energy demand in developing Asia will double over the next 25 years,[18] the majority of which stems from economic and population growth in developing countries and transitional countries.[19]

The demand for development and industrialization, and their incumbent consumer benefits, is virtually insatiable in the intermediate term. Some of Asia's *miracle* nations, including the so-called *tigers*, have illustrated that it is possible within one generation to accomplish levels of industrial development that took other nations a century or more to achieve. Moreover, the liberalized trade policies of most nations and the goals and policies of international organizations foresee rapid development and electrification of less-developed nations as part of an increasingly interlinked globalized economy.

Access to modern electric energy is a fundamental part of human development and an important catalyst for economic growth, especially in developing nations. Increased access ultimately results in access to necessities and an improved standard of living. Direct benefits from electric energy access include reliable lighting, heat and refrigeration, health benefits because of enhanced indoor air quality from the cessation of use of polluting fuel sources, reduced fire hazards, higher-quality health services equipment, greater business productivity, and increased opportunities for education.[20]

To illustrate the discrepancy between energy use and standard of living, the United Nations Development Programme (UNDP) developed a human development index (HDI) for comparing social and economic well-being in relation to per-capita energy use. The HDI values, shown in table 1–1, illustrate a strong correlation between social and economic well-being and per capita energy consumption levels. Countries with high HDI values have high per capita energy consumption values. Conversely, countries with low HDI values, typically developing countries, have low per capita energy consumption values.

Table 1–1: Per Capita Energy Use By Region 2000[1]

Asia	China	Former USSR	OECD Europe	OECD North	OECD Pacific America	Latin America/ Caribbean	Sub-Saharan Africa Caribbean
25	38	133	142	281	180	48	25
Source: 2004 World Energy Assessment, at 27							

The HDI values for most of the population in developing countries is distorted or inflated because of the maldistribution of modern energy services in those locations. In fact, there is a significant energy consumption dichotomy within developing countries when contrasting the energy services consumed by affluent subgroups of the population with those available to low-income households. The affluent portion of the population generally represents a small subsection of the overall country population but disproportionately consumes energy.[21] This division reflects the significant stratification of incomes in developing nations.

Greater than 70% of the developing world still relies on wood or charcoal as a primary fuel source for cooking and heating.[22] UNDP calculations indicate that biomass fuels account for approximately 25% of all energy generation in Asia.[23] There are multiple health and resource disadvantages, however, to the use of biomass fuels (although electricity will not displace most biofuels because cooking with electricity is unlikely in developing nations):

- Gathering wood for fuel in developing nations consumes much of the time of women and children; access to modern fuels greatly reduces this time, enabling them to spend this time on education and leisure.

- The UNDP estimates that approximately 1.6 million people, predominately women and children, die prematurely each year due to particulate pollution from indoor burning of biomass fuels.[24] In India alone, there are more than half-a-million deaths each year attributable to indoor air pollution from the burning of traditional fuels.[25]

Technology. The only one of these three variables in the global warming equation that can be influenced dramatically by policy makers is the choice of development technology. The basic touchstone technology for development in all nations is electricity, which has been described as an "agent of technological progress." Because electricity is used in place of fossil fuels and human labor, less overall energy is necessary, and more productive and efficient operations occur in certain segments of society.[26] Yet, electricity is so indispensable that it has become "transparent to most users, at least until there is an outage."[27]

There is good news and bad news concerning the choice of technology. The good news, as elaborated later in this chapter as well as in the next, is that today there is a variety of proven and reliable renewable energy technologies to supply electric power available that do not contribute significantly to global warming. Many developing nations have a single, centralized state-owned utility that monopolizes the nation's electricity supplies. In this framework, a single decision about the electric power development path could be implemented nationwide.

The bad news is that not all renewable technologies are yet cost competitive in many applications. In many rural applications of the existing transmission grid, however, it is more cost effective to install a dispersed renewable energy technology to provide electricity than it is to extend the transmission grid to the region to supply centrally generated electricity. But in other situations, not accounting for the environmental benefits of renewable electricity, these technologies can be dear. Their costs typically are incurred up front, rather than over the life of the electric generating unit, as is common with fossil fuel-fired facilities, which must purchase fossil fuels over their operating lifetime. In addition, many developing nations have large supplies of coal; the temptation to burn cheap coal in conventional electric power production first—as in the United States, the United Kingdom (UK), and elsewhere during their industrial revolutions— is palpable. It is of note that during the industrial revolutions in the United States and UK in the 18th and 19th centuries, coal was the only energy technology available at a scale to fuel manufacturing industry, with mechanical power from hydro facilities available in certain places. Oil, natural gas, and other renewable technologies were not then available as they are today.

The science of global warming

Historically, climate change is important. There is historical evidence that societal collapse has occurred from global climate change.[28] Climate change is attributed to global warming because of the greenhouse effect from anthropogenic emissions of CO_2 and other GHGs.[29] Like the greenhouse effect itself, climate change is a naturally occurring process, but it has been intensified by human activities. The risks of climate change include the potential for large-scale and possibly irreversible impacts on continental and global scales.[30]

The greenhouse gases in our atmosphere

Greenhouse gases are a direct byproduct of the kind of economic growth Asia is experiencing. GHGs occur naturally in the atmosphere, but since the industrial revolution, emissions resulting from combusting fossil fuels for mechanical and electrical energy have poured into the atmosphere. GHGs trap sunlight in the earth's atmosphere, absorbing the longer infrared radiation, and turn the sunlight into heat, a phenomenon known as the *greenhouse effect*.

Greenhouse gases include as those gases of most concern such as carbon dioxide (CO_2), methane (CH_4), nitrogen oxides (NO_x), nitrous oxide (N_2O), sulfur dioxide (SO_2), volatile organic compounds (VOC), Ozone (O3), particulate matter (PM-10, 2.5), sulfur hexafluoride (SF_6), hydrofluorocarbons (HFCs), and perfluorocarbons (PFCs). All are measured in CO_2 equivalents, CO_2e. The Kyoto Protocol regulates emissions in developed countries of these greenhouse gases.

The secondary greenhouse gases, resulting significantly from agriculture, electricity production, and transportation, all contribute to and intensify the greenhouse effect with varying amounts and levels of potency. In 2000 anthropogenic activities emitted 320 million tons of methane and 33 million metric tons (TgN) of NO_x into the atmosphere per year; these levels are rising at a rate of about 4% per year.[31] These secondary GHGs are more highly reactive in the atmosphere, and while they have local environmental impact on air quality, their impact on the greenhouse effect is not completely understood.[32]

The global warming impact molecule-by-molecule of many of these secondary, and less prevalent GHGs, is significantly greater than CO_2. However, because they are released in much lesser quantities and/or have shorter residence times in the atmosphere before they dissipate, CO_2 is the most troubling GHG. Assigning CO_2 a global warming potential of 1, the relatively greater magnitude of the other GHGs and their residence time in the atmosphere is as provided in table 1–2. The GHGs in table 1–2 are displayed in descending order of their impacts on the environment, which is a function of their quantity released, heat radiation properties, and residence time in the atmosphere.

Whatever the impacts of the secondary GHGs, CO_2 is far and away the largest GHG by emitted volume. Carbon dioxide is the main byproduct of fossil fuel combustion, and therefore results from any energy production that uses oil, coal, natural gas, or other solid waste fuels. Ninety-eight percent of anthropogenic CO_2 emissions are from combustion of fossil fuels, and 83% of U.S. GHG emissions are attributed to CO_2.[33]

Most of the world's carbon is collected in the deposits of oil, gas, and inorganic deposits of carbonates. Carbon is also in plants in the ocean and growing plants and trees on land. Most organisms use organic molecules or carbon for the energy required for growth. Approximately 15 billion tons of carbon, in the form of CO_2, is turned into new wood each year.[34] Forests are very important in the CO_2 cycle, containing roughly two-thirds of atmospheric carbon.

Table 1–2: Key Facts about Greenhouse Gases

GHG	Global Warming Potential [CO_2 = 1]	Residency Time [years]	Amount of U.S. Total GHG Release [%][1]
CO_2	1	100	85
Methane	21	12	11
Nitrous Oxides	310	120	2
Hydrochlorofluorocarbons	140–11,700	varies	< 1
Chlorofluorocarbons	6,500	varies	< 1
Hexafluoride (SF_6)	23,900	varies	< 1

[1]U.S. EPA, *Inventory of U.S. Greenhouse Gas Emissions and Sinks: 1990–1993*, EPA 230-R-94-014 (1994), at ES-2.

Because CO_2 lacks the immediate environmental impacts of NO_x or SO_2, the U.S. Environmental Protection Agency does not classify CO_2 as a *pollutant*. But the sheer amount of CO_2 emitted into the environment is enormous and persists for more than 100 years. In 2000 the world emitted almost 7 billion tons of CO_2 into the atmosphere per year.[35] Global CO_2 emissions are rising at the rate of approximately 10% per year.[36] Atmospheric CO_2 levels now are approximately 33% higher than in preindustrial times.[37]

Almost three-quarters of anthropogenic CO_2 emissions are generated in developed countries, although this balance is shifting toward developing countries.[38] Although emitting less than one-quarter of anthropogenic sources of CO_2 now, developing nations are expected to emit a majority of CO_2 emissions before 2035, whereas China is expected to surpass the United States as the largest CO_2 emitter in the world within 10 years.[39] The sobering news in this is that not only are emissions increasing, but developing countries have much higher GHG emissions per unit of gross national product.[40] China has the highest emissions in the world per unit of gross national product (GNP) by a factor more than double other nations; as these nations industrialize, they increase GHG emissions; although CO_2 intensities may decline as the economies grow more efficient, and economic activity shifts becomes less energy intensive, as has been observed in developed countries.

The two main sources of anthropogenic GHG emissions are motor vehicles and electricity generation plants.[41] Asia's economic growth, high population growth, and increased urbanization are increasing pressure on both transportation and electricity generation resources. In Beijing during 2003, vehicle ownership rose (from 1 million in 1997) to 2 million, and new vehicles came onto Beijing's roads at a rate of 27,000 per month. In many cities in India, the dominant motor vehicles are "auto-rickshaws," vehicles with small two and four stroke engines that use a highly polluting home-brew of a kerosene-lubricant mix as fuel.[42] Implementing limits on personal transportation is a daunting challenge. Unlike electricity, which is often centrally supplied from a few large plants, transportation decisions on vehicle purchase and use are made by billions of individual households in an atomized and disaggregated fashion. The extremely low capital investment needed to get a motor vehicle on the road seriously impedes emissions control. To affect consumption decisions in the transportation sector requires motivating hundreds of millions of individual consumers. Similarly, the extremely low capital intensity of an electricity using appliance is an obstacle to CO_2 emissions reductions.

Electricity generation is the other major producer of GHGs; fossil fuel-fired power plants are responsible for 30% of anthropogenic CO_2 emissions.[43] China currently meets 70% of its electricity demand through coal plants, the most prolific emitters among fossil fuel plants in terms of both CO_2 and particulate matter; 57% of India's electricity comes from coal.[44] To meet its growing energy needs, China intends to roll out 100 new large power plants by 2020, including nuclear, hydropower, and coal plants.[45] India has targeted 100,000 MW in new capacity over the next ten years.[46] Vietnam plans to add scores of additional new hydroelectric and oil-fired plants by 2010.[47] The single-point nature of power plants' emissions, the centralized nature of most power plant decisions in developing nations, and the exploding demand for electricity, make electricity-generating plants the logical choice for a frontal assault on GHG emissions.

The case supporting long-term global climate change

The Intergovernmental Panel on Climate Change (IPCC), a UN agency created during the 1990 Rio de Janeiro Earth Summit to serve as a non-politicized source of information about climate change for policy makers, has three working groups. In 2001 the IPCC's three working groups published their most recent major reports, along with summaries for policymakers, which contain the distilled findings of the working groups.

The nearly 7 billion tons of CO_2 emitted into the atmosphere annually stays in the atmosphere for extremely long periods—100 years—and accumulates over time. The IPCC estimates that the present atmospheric concentration of CO_2 is at its highest level in the past 420,000 years, and the current rate of increase is unprecedented over the past 20,000 years.[48] The result of this massive influx of carbon into the atmosphere is a rapidly warming climate.[49] During the 20th century, the global average surface temperature increased 0.6°C—likely the northern hemisphere's warmest in a thousand years. By 2100 IPCC models project the average global surface temperature to warm anywhere from 1.4°C to 5.8°C.[50] This is a rate of warming higher than has occurred over the past 10,000 years. The IPCC concludes that it is very unlikely that such warming is natural in origin or because of internal variability alone.

The IPCC uses global mean annual temperature as a proxy for the magnitude of climate change, but projected impacts depend on multiple factors, such as the magnitude and rate of global and regional variation in mean climate, extreme climate phenomena, socioeconomic conditions, and adaptation.[51] Flooding and sea level rise resulting from climate change pose the most direct risk to human settlement. In developing nations, the impacts of climate change are expected to be far reaching, adversely affecting virtually all aspects of social and economic life, threatening agriculture and water supply, and displacing millions of people living in low-lying areas.

In some of the IPCC's projections, the average sea level will rise anywhere from 0.09 to 0.88 meters, displace 200 million people by 2080, and cause tens of billions of dollars of damage to coastal-area infrastructure.[52] "Continued global warming is in nobody's interest, but the simple facts of the matter are that developing countries will suffer the most damage, and their poor will be at an even greater disadvantage."[53] Human systems are also vulnerable to extreme climatic events, which will increase in severity with continued global warming. The economic loss from ordinary and extreme weather events has soared in recent decades. Global economic losses from extreme weather events rose by a factor of 10.3 from $3.9 billion per year in the 1950s to $40 billion per year in the 1990s.[54] The costs of such events have increased rapidly despite significant efforts to fortify infrastructure and enhance disaster preparedness.

Finally, the impacts of climate change may not have a linear relationship to increased temperature or GHG concentrations.[55] As a complex biological system, the climate exhibits complex, nonlinear behavior, which may resemble cumulative chaotic scenarios.[56] In other words, there may be a *tipping point* or precipice at which naturally balanced systems spiral rapidly into disarray. Because we have never advanced to such a precipice, there is no empirical demonstration of where or whether it exists.

The real danger lies in the small, but serious potential for a catastrophic event with unpredictable consequences. Examples of such extreme impacts include the slowing of the Gulf Stream current and large reductions in the Greenland and West Antarctic Ice Sheets.[57] If such changes do occur, their impact could be widespread, sustained, and irreversible in any discreet time frame in which human political institutions can negotiate and respond.

The dissenting case contradicting long-term global climate change

Although the IPCC acts as the voice of scientific consensus, there are numerous scientists who dissent from the IPCC's conclusions. The dissent proceeds on several grounds:

- It contests the scientific data and its interpretation.

- It argues that additional CO_2 emissions will lead to additional plant carbon sink development in the tree canopy and oceans, or other natural balancing factors, which will absorb or neutralize the additional CO_2.

- It submits that future technological innovation will mitigate any global warming impacts.
- It attacks the phrasing of the IPCC conclusions.[58]

The dissent in the scientific community submits that given the climate's natural variability and of the IPCC's own scenarios, there is no compelling evidence that global warming is anthropogenic.[59] Massachusetts Institute of Technology (MIT) meteorologist Dr. James Lindzen has attacked the methodology of the IPCC's historic temperature analysis. Attacking the claim that the last century was the warmest in 1,000 years, Dr. Lindzen argues that the IPCC researchers used tree rings alone to gauge temperature for the first 600 years of the study, from only four separate locations. Lindzen calls the method to turn tree-ring width into temperature "hopelessly flawed"[60] and claims that the IPCC seriously overstates the correlation between temperature and the amount of CO_2 in the atmosphere. Additionally, critics of the IPCC point to a 1,000 year climate study by Dr. Sallie Baliunas and Dr. Willie Soon of the Harvard-Smithsonian Center for Astrophysics.[61] Soon and Baliunas take issue with the IPCC by contending that the 20th century saw no unique patterns; they found few climatic anomalies in the proxy records.

In 2000 James Hansen, the NASA scientist credited with initially raising the alarm about global warming, concluded that the net carbon impact of fossil fuel combustion has been close to net zero because aerosol cooling offsets CO_2 warming. He concluded that the world community should shift its focus and concern from CO_2 to the other GHGs. Such a shift would focus less on reduction of fossil fuel use in power plants.

Senator James Inhofe, chairman of the Senate Committee on Environment and Public Works, has specifically criticized the wording of the IPCC Summary Reports, saying that "some parts of the IPCC process resembled a Soviet-style trial, in which the facts are predetermined, and ideological purity trumps technical and scientific rigor."[62] Inhofe argues that the wording of the reports is designed to "maximize the fear factor," and further argues that the extreme warming scenario forecast of a rise of 5.8°C is "ludicrous" because it is based on the assumption that energy production will remain carbon intensive.[63]

Institutional responses to scientific uncertainty

Kyoto. How do institutions and nations respond in the face of scientific uncertainty? They have responded without substantial unanimity. More than 30 years ago, the U.S. Council on Environmental Quality issued this call for an intergenerational approach to the possibility of global warming:

> One imperative we share is to protect the integrity of our fragile craft and the security of its passengers for the duration of our voyage. With our limited knowledge of its workings, we should not experiment with its great systems in a way that imposes unknown and potentially large risks on future generations. In particular, we cannot presume that, in order to decide whether to proceed with the carbon dioxide experiment, we can accurately assess the long-term costs and benefits of unprecedented changes in global climate....Although our domination over the earth may be nearly absolute, our right to exercise it is not.[64]

By 1991 13 developed nations, not including the United States, had agreed to reduce or stabilize their CO_2 emissions by 2005. These pledges have not been realized. The Framework Convention on Climate Change treaty was accepted at the Rio de Janeiro UN Conference on Environment and Development in 1992 and the Kyoto convention in 1997. The Kyoto Protocol requires 38 developed nation by 2000 voluntarily to reduce CO_2 emissions, and by 2012 to reduce CO_2 emissions 7% below 1990 baseline levels. The other GHGs must be reduced to 5% to 7% below either their 1990 or 1995 baseline levels by 2008 to 2012. Emissions may be reduced or forest canopy expanded to absorb CO_2.

Many nations signed the Kyoto Protocol, but they were slow to ratify and thus implement it. The United States has withdrawn from the Kyoto Protocol, whereas most European Union (EU) countries, including Russia, have ratified.[65] To advance the goal, the EU has already established an internal target of 22% renewable energy for the generation of electricity and 12% of all energy from renewable resources by 2010, although it may not be realized.[66]

Only 38 of the 200 world nations are covered in the Kyoto Protocol by mandatory obligations. Because of the short-term needs of developing countries to provide food, energy, and other services to their poor, they are not required by the Kyoto Protocol to invest in GHG reductions. In fact, there is no mechanism in the Kyoto Protocol to ensure compliance of any nation; it is expected that developed nations will engage in concessional technology transfer and assistance with renewable energy technologies in developing nations. An international treaty established the Global Environmental Facility (GEF) in 1991 and later established the Carbon Finance Business facilities to assist with renewable energy development in developing nations (see chapter 10 for a more detailed discussion of these facilities and financing). Policy reforms are necessary in many of the developing nations to accommodate a vigorous market-based renewable energy program with a level playing field, as examined in chapters 3 to 9.

These targets for GHG reduction will not be achieved in the specified time frames. The U.S. Department of Energy (DOE) forecast that a worldwide increase of 54% more GHG than 1990 levels could occur by 2015.[67] Although greenhouse gases in the United States since 1990 have increased more slowly than population growth or electric power production, in the one dozen years after 1990, U.S. greenhouse gas emissions increased 10.9%.[68]

Common but differentiated responsibility

The United Nations Framework Convention on Climate Change (UNFCCC) uses the principle of *common but differentiated responsibility* with regard to climate change. Much of the international community views GHG as a common but differentiated responsibility of both developed and developing nations.

> *Common but Differentiated Responsibility articulated as Principle 7 of the Rio Declaration, this principle requires states to cooperate in a spirit of global partnership to protect the environment. Yet, because states have contributed differently to global environmental problems, the principle recognizes that they should have common, but differentiated, responsibilities. A good example is Article 4 of the 1992 UNFCCC, which places an obligation on developed countries to take the lead in meeting the required reductions in greenhouse gas emissions. Developing country parties, however, are only obliged to implement these commitments to the extent that developed countries have met their commitments to provide financial resources and to transfer technology. As a general principle, sure to govern further negotiations on the UNFCCC, the principle of common but differentiated responsibility is highly significant. The structure of the 1997 UNFCCC Kyoto Protocol mirrors the philosophy of common but differentiated responsibility. Developed countries are committed to*

> reducing their overall emissions of greenhouse gases by at least 5% below 1990 levels between 2008 and 2012. Developing nations have no such commitments. Although every nation state has the responsibility to reduce global greenhouse gas emissions, only Organisation for Economic Co-Operation and Development (OECD) and economies-in-transition countries are required to make specific, quantified emission limitations. The limitations, even among these countries, vary to take into account differing domestic circumstances. Developing countries are provided with an opportunity to participate through the Clean Development Mechanism, which allows countries to cooperate on specific projects to reduce greenhouse gas emissions.[69]

The developed nations have much larger absolute and per capita energy consumption but in some instances lower energy consumption per unit of gross domestic product. With population more stable in many developed nations, these are the locations where population and development pressures will lead to dramatic increases in electric generation in the next two decades.

The Kyoto Protocol imposes GHG reduction obligations only on developed nations. The developing countries are vested with responsibility to control GHGs but no requirement. This leaves a substantial responsibility on developed nations to transfer technology on beneficial terms and assistance to help developing nations with these matters.

Two Scientific Predictions, One Path

The second question is not how institutions and nations *have* responded, but how they *should* respond. On this question, the lack of scientific experimental evidence or validation should not create institutional inertia. Ultimately, we do not know with certainty what impact GHGs will have on the planet over time. Whether conventional fossil fuel–fired power plants are (1) permanently torquing the global thermostat beyond natural limits of sustainability or (2) merely expelling copious quantities of GHGs and other criteria pollutants into the atmosphere, the prudent policy response for developed and developing nations may be the same. There is a logical policy choice

- when sustainable power generation technologies are available which do not cause significant GHC emissions

- at an appropriate scale and application to the demands of developing nations;

- which nations have a flexible and developing power infrastructure than can accommodate *greenfield* renewable energy projects
- where developed nations and international organizations are prepared to provide substantial financial and technical assistance with alternative energy development for developing countries
- deployment of renewable resources buffers developing countries particularly, and all nations generally, against the volatile financial and supply vicissitudes of importing fossil fuels for their power sectors

Under either global warming scenario, it makes logical institutional and national sense to deploy renewable resources to a significant degree. As discussed in the next chapter, many renewable technologies make economic sense now in terms of economics and buffering users from fossil fuel–price fluctuations, even without taking account of GHG issues. Renewable energy technologies in many instances—particularly where new electric supply infrastructure is being created or extended—are justified without regard to their GHG benefits in developing nations.

If the forecasts for dire global warming impacts are correct, nations and the international community will wish it had done *more, earlier* than current efforts to deploy renewable power technologies. If, to the contrary, the dire GHG predictions are not borne out over time, the worst-case scenario is that developing and other nations, for some part of their power mix, have elected proven renewable energy technologies in lieu of some conventional technologies. The IEA in Paris forecasts that by 2030, world demand for energy will grow by 59% and fossil fuel sources will still supply 82% of the total with noncarbon renewable energy sources supplying only 6%.[70]

The World Bank Group has funded more than $6.3 billion in renewable energy projects since 1990.[71] If the international community underwrites some of the added first capital costs of these renewable energy technologies, are developing nations somehow disadvantaged if global warming does not materialize? The sheer dollar cost of fossil fuels will rise as demand for electricity grows and fossil fuel becomes more scarce and harder to obtain.[72] A supply-side increase in the rate of extraction of fossil fuels will not indefinitely solve the fossil energy supply problem. Fossil fuels are created during more than hundreds of millions of years, and thus in modern institutional terms, are finite and not renewable. Doubling the size of world oil reserves will add, at most, 14 years to the life expectancy of the reserves if use continues to climb at the currently increasing rate.[73] It is generally acknowledged that because of reasons of dwindling accessible supply and price, the voracious energy appetites of humankind will cause a shift to alternative energy sources. This inevitability presents a technological and economic advantage for whichever nations build power infrastructure now, at least in part, around noncarbon fuels.

A nation that begins implementing that shift now, rather than waits until economic or physical realities compel a sudden shift away from fossil generation, is not necessarily worse off. Renewable energy alternatives, which are not tethered to the supply of imported fossil fuels to operate, guarantee those implementing nations some energy independence. For those nations, it is an insurance policy against the economic, lifestyle, and political upheavals that characterize power supply disruptions. For all nations, diversification of generating technologies is a cornerstone of sound energy policy, national energy security, and policy regarding global warming.

The Precautionary Principle and Scientific Uncertainty

Whatever the ultimate outcome of the scientific debate on global warming, lack of certainty surrounding the possible effects of climate change is not a good rationale for inaction. Estimates of the benefits of reducing GHGs range from $5 per ton to $125 per ton, with an additional benefit of up to $20 per ton for diminution of criteria pollutants.[74] The enormity and complexity of climate change means that the policy-making time frame is very long, and policy makers will be unable to affect a quick fix.[75]

The international legal doctrine known as the *precautionary principle* advances the concept of policy risk hedging to respond to ecological uncertainties; it works to keep options open for the future.[76] The precautionary principle militates in favor of limiting GHGs to a level that eliminates any foreseeable plausible threat of catastrophic environmental scenarios. The UNFCCC, still the only international accord on climate change in force, makes the precautionary principle a core tenet in Article 3:

> The Parties should take precautionary measures to anticipate, prevent, or minimize the causes of climate change and mitigate its adverse effects. Where there are threats of serious or irreversible damage, lack of full scientific certainty should not be used as a reason for postponing such measures, taking into account that policies and measures to deal with climate change should be cost-effective so as to ensure global benefits at the lowest possible cost.[77]

The precautionary principle addresses scientific uncertainty by implementing policy alternatives that minimize the maximum downside risk. It is the policy equivalent of an insurance policy. Given the scientific uncertainty surrounding the future impact of global warming, it implements alternatives now to guard against the more disruptive forecast possibilities.

The Total Cost of Fossil Fuel Dependence

Beyond the precautionary principle, the UNFCCC's emphasis on cost-effective *policies and measures* with *global benefits* portends the significant costs of vastly increased fossil fuel use, and significant benefits of substituting, to some significant degree, renewable energy. There also are compelling economic reasons to reduce fossil fuel use and develop renewable energy sources. According to some observers, the massive military investment of the United States in the Middle East, and resulting political instability, is part of the cost of dependence on foreign fossil fuels.[78] Over the past three decades, American consumers have transferred $7 trillion in wealth to the Organization of the Petroleum Exporting Countries (OPEC) nations in return for oil.[79] By contrast, renewable resources transfer that money not to foreign oil suppliers, but to manufacturers and suppliers of renewable energy equipment, which increasingly have the potential to be domestic suppliers.

There are other environmental costs from burning fossil fuels besides possible climate change and global warming. For example, the environmental costs associated with human health and land degradation as a result of fossil fuel use are large enough to overshadow China's remarkable economic expansion.[80] Using a conservative estimate, the U.S. embassy in Beijing calculated that environmental degradation is costing China 7% to 8% of its annual GDP, a number greater than China's annual economic growth rate.[81]

In contrast, renewable resources are abundant and diverse. As indigenous resources, they encourage both local control and economic growth. Renewable energy sources provide an important hedge against rising fossil fuel prices. In deregulated electricity markets, renewables reduce energy costs for the entire power market because they lower the marginal clearing price of electricity. The marginal cost of renewable fuel sources is essentially zero, and renewable energy alternatives have substantially less negative environmental impacts than fossil fuel sources over the long term.[82] Many renewable energy facilities can be built in size increments proportionate to load growth patterns and local needs. Furthermore, smaller plants located closer to the customer load reduce infrastructure costs for T&D, reduce losses, and help secure local power reliability and quality.

The Asia dimension

The war on global warming cannot be won without success in Asia. Although the industrialized West currently emits the lion's share of greenhouse gases, Asia's rapidly expanding economies, combined with high populations with growing wealth and a demand for an energy-intensive Western lifestyle, mean that Asia is on course to vastly increase the level of global GHG emissions if it chooses conventional power generation technologies. China is already the world's second leading emitter of GHGs, and such emissions in India and Indonesia threaten to skyrocket.[83] The U.S. DOE projects that energy demand in developing Asia will double over the next 25 years.[84] The IEA in Paris forecasts that two-thirds of all future energy demand will emanate from just China and India. Some projections estimate that by 2030 China's GHG emissions will quadruple, and Asia alone will emit 60% of the world's carbon emissions.[85]

This future is directly dependent on whether fossil fuels or renewable technologies are chosen now to generate electricity. This is no small choice. It is expected that global energy use will double by 2040 and triple by 2060, creating a tremendous demand on existing fuel sources.[86] There is a policy choice involved between conventional and alternative resources.

There are enough conventional fossil resources in the intermediate term to make this a real choice. Resource economists believe that Asia has fossil fuel reserves enough to last for longer than 100 years. More than 90% of these fossil reserves are coal, however, and several of these nations, most notably China, are already highly dependent on coal as their principal energy source.[87] In 2003 alone China's oil consumption jumped by nearly one-third, domestic coal production increased by 100 million tons, and electricity consumption rose by 15%.[88]

In response Chinese total installed electric generation capacity grew from 65 GW in 1980 to 353 GW in 2002, making it the second largest in the world after the United States.[89] Electricity demand between 1996 and 2000 grew at an average of 6.3% annually and is expected to almost match this pace into the future.[90] To avoid shortages and satisfy demand, China would have to increase electric capacity by approximately 40 GW annually.[91] Of this capacity addition, at the Bonn International Conference on Renewable Energies in June 2004, China committed to adding approximately 6 GW of renewable energy annually until 2020.[92] The funding now of renewable energy projects worldwide, and especially in Asia, is necessary to prevent these nations from becoming even more reliant on a fossil fuel–based generation infrastructure. The IEA projected that it will require an investment of $16 trillion by 2030 to meet the world's energy requirements, with $5 trillion of that amount allocated to electric power production, primarily in Asia and Africa.[93]

Underlying Asia's burgeoning energy demand is a low per capita energy use base today, growing urbanization, and high consumer demand for Western style amenities. Leaving its rural roots, Asia is migrating rapidly toward an urban future, with rapidly growing per capita energy consumption. By 2025 one quarter of the world's population will be living in Asian cities.[94] Urbanization and population growth in India have driven a 208% growth in India's energy consumption in the last 20 years.[95] In India, despite harsh economic conditions where a quarter of the population makes an income of less than a dollar a day, sales of consumer goods from shoes to shampoos are rising sharply.[96] In Indonesia sales of cars and motorbikes are growing annually anywhere from 10% to 20% on average.[97] China is the world's second largest energy consumer behind the United States, but on a per capita basis, the Chinese consume only one-tenth of the energy Americans use.[98]

The intimate link between fossil fuel consumption and macro-economic growth is a major factor in Asia's potentially explosive contribution to global GHG emissions. The economies of China and India are growing at a rapid pace, with projected annual growth rates of almost 6% for the next quarter century.[99] China, India, Indonesia, Thailand, and other Asian nations are already major manufacturing centers, and capital investment from the West has flowed to all sectors of their economies as trade barriers have fallen. India's private sector, buoyed by a growing consumer market and the outsourcing of technically skilled jobs from the United States, is demanding electric infrastructure growth and reliability improvements from the Indian government. These examples are the stories of all developing nations, made larger by the vigor and numbers of the Asian economies.

The needs of countries outside the Organisation for Economic Cooperation and Development (OECD) will require an investment of some $2 trillion to install approximately 1900 GW of new electric generating capacity by 2025.[100] Of a population in year 2000 of just more than 6 billion persons, East Asia and China contained more than 1.8 billion persons and South Asia contained more than 1.3 billion persons, collectively constituting in Asia more than half the world's population.[101] To be successful in any environmental battle on global warming, the sheer numbers of population coupled with rapid development, the challenge must be met successfully in Asia for the world to succeed.

Infrastructure Lifetimes and Hard Choices

The choices for many developing countries are challenging. We stand at a crossroads; in the next two decades, there will be a massive electrification of developing nations when developing nations will choose whether to deploy conventional fossil-fired or sustainable renewable options to generate electricity. Once installed, those facilities will remain in place, contributing to global warming or not, often for 40 years and in many cases longer.

These are *hard* infrastructure choices. Experience in the United States demonstrates that older fossil-fired power plants, at the conclusion of their originally scheduled lives, typically are refitted with new burners, boilers, and fuel-handling equipment and extended for additional decades. For the past decade in the United States, there has brewed a pitched battle between environmentalists and utilities over the life-extension of fossil-fired power plants. As these plants have reached their scheduled life, they often have been refitted with new equipment and kept in service. A battle still bounces from the regulators to the courts about whether this is allowed without meeting more stringent Clean Air Act requirements for New Source Review.[102] Moreover, once the transmission infrastructure is established to carry power out of a large fossil-fired plant to load centers, this creates a transmission and distribution (T&D) corridor, system hardware and distribution patterns that require a centralized large power generation facility at that terminus of the transmission grid.

Electric T&D facilities, telecommunications equipment, and oil and gas pipelines have long lives.[103] Once a T&D system is created to link centralized generation with distribution, it becomes an embedded *hard* infrastructure. This is where distributed generation and renewable technologies may offer some accommodation; they either can be placed on site at existing centralized generation locations, or distributed solar technologies can be sited at many dispersed locations.

Like a highway grid, once configured, locational and use patterns that grow up around that grid make it more difficult to reroute those electric highways. *Hard* infrastructure choices of any kind, once embedded in the physical and distributional fabric of a country, are not easily removed or altered. This is not to say that one cannot later substitute in place a fossil-fired unit which has reached the end of its useful life with a renewable unit, but it is often practically impossible. Conventional fossil-fired projects typically have been sited either: (1) where fuel supply, transmission off-take capacity, and cooling water resources coincide; or (2) in transmission proximity to population and load centers.[104]

Renewable technologies must go to the place where they can be exploited. Only in certain locations is the wind regime sufficient to turn large wind turbines; hydro power is limited to moving water courses; solar photovoltaic power, while ubiquitous, requires a large land or surface area to produce the equivalent amount of power as a large fossil fuel-fired facility (solar power is much less dense than fossil fuels—although solar collectors can be mounted of roofs or walls, or have dual uses, e.g., functioning as both a roof and electricity generator).[105] Therefore, it does not follow that older fossil-fired facilities can or will be replaced at their sites with renewable power technologies. With a mature T&D system in place, the total system economics and technical considerations may militate in favor of continuing the existing fossil fuel facilities.

Therefore, without diminishing the importance of renewable technologies gradually replacing existing but obsolete fossil fuel-fired facilities in developed systems, the critical challenge is what is deployed in developing nations to meet rising demand and extension of service to previously unserved areas. This is where there is population growth, pressure for rapid electrification, and a developing infrastructure that can accommodate either renewable or conventional technologies:

> *Developing countries offer unique opportunities for cultivating sustainable energy in large part because the bulk of their energy demand and investments still lie before them.... The World Bank Group is committed to nothing less than a evolution in the rate and scale with which sustainable clean energy services are expanded to those who lack them, and the new dimension in global partnerships that is needed to bridge the modern energy divide.*[106]

There is no expectation that any nation will deploy exclusively fossil fuel or only renewable electric generation technologies. It is not an *either/or* choice. However, the balance chosen between conventional and alternative electric resources has immense implications for the emission of greenhouse gases. The critical path timing of these decisions is now. And Asia, with 60% of forecast future world electrification in this one continent alone, is the key ground on which this battle to limit GHGs must be won. Although decisions in all developing nations are important, one cannot lose the battle against global warming in Asia, and make up the loss in other nations. Therefore, Asia is the laboratory, and electricity is the crucible, which will determine the global warming mix.

Notes

1. Fritsche, U., and F. C. Matthes. 2003. Changing course: A contribution to a global energy strategy. Presented at the 2003 World Summit of the Heinrich Boll Foundation, Johannesburg, South Africa, Paper No. 22; 13.

2. Ibid., 15.

3. Ibid., utilizing IEA 2002 data.

4. Barnes, D. F., and J. Halpern. 2000. The role of energy subsidies. *Energy services for the world's poor: Energy and development report 2000*, 60. Washington, DC: The World Bank, Energy Sector Management Assistance Programme.

5. The World Bank Group. 2004. *2004 World energy assessment*, 33. Washington, DC: World Bank. (estimating 2 billion people); Brook, P. J., and J. Besant-Jones. 2000. Reaching the poor in the age of energy reform. *Energy services for the world's poor: Energy and development report 2000*, 2. Washington, DC: The World Bank, Energy Sector Management Assistance Programme. (estimating 2 billion); Dubash, N. K. 2002. *Power politics: Equity and environment in electricity reform*, 2. Washington DC: World Resources Institute. (1.7 billion); Energy and Mining Sector Board. 2001. *The World Bank Group's energy program: poverty alleviation, sustainability, and selectivity*, 7. Washington, DC: The World Bank Group. (1.6 billion).

6. Dubash, N. K. 2002.

7. Goldemberg, J., and T. B. Johansson. 2004. *World energy assessment: Overview 2004 update*, 65. New York: United Nations Development Programme.

8. International Energy Agency. 2003. *World energy investment outlook.* Paris: International Energy Agency.

9. Ibid. Almost one-third of the people in the world have no access to electricity, and many of those who have access do not consume it for economic and financial reasons.

10. Price, C. W. 2000. Better energy services, better energy sectors—and links with the poor. *Energy services for the world's poor: Energy and development report 2000*, 28. Washington, DC: The World Bank and Energy Sector Management Assistance Programme; Townsend, A. 2000. Energy access, energy demand, and the information deficit. *Energy services for the world's poor: Energy and development report 2000*.

11. Goldemberg and Johansson, 2004. (pg. 26–27).

12. World Bank Group. *Renewable energy for development*, 4. Washington, DC: World Bank Group.

13. Ibid., 18.

14. Goldemberg and Johansson, 2004 (pg. 13).

15. Brook and Besant-Jones, 2000 (pg. 2); Ibid., 2 (starting at $50); Price, 2000 (pg. 28). ($80–$300); Townsend, 2000 (pg. 12). ($50–$1000).

16. Goldemberg and Johansson, 2004 (pg. 31).

17. International Energy Agency. 2002. *World energy outlook 2002*. Paris: International Energy Agency

18. International Energy Agency. 2004. *World energy outlook 2004*. Paris: International Energy Agency, http:// www.worldenergyoutlook.org

19. *Renewables 2004*, 7. Conference issue paper at the International Conference for Renewable Energies, Bonn, Germany, June 1–4, 2004.

20. Price, 2000. (pg. 26–27)

21. Goldemberg and Johansson, 2004, (pg. 26).

22. Ferrey, Steven. 2005. *The law of independent power*, 22nd ed., sect. 3:19. St. Paul, MN: West Publishing.

23. Goldemberg and Johansson, 2004 (pg. 30).

24. Ibid., 40.

25. Energy and Mining Sector Board, The World Bank Group. 2001. Topical briefing to the board of directors on energy. *The World Bank Group's energy program: Poverty alleviation, sustainability, and selectivity,* 12. Washington, DC: World Bank Group.

26. Gellings, C., and R. Lordan. 2004. The power delivery system of the future. *Electricity Daily* (January–February).

27. Ibid.

28. Weiss, H., and R. S. Bradley. 2001. What drives societal collapse? *Science* (January 26): 98.

29. Pew Center for Climate Change. Global warming basics. http://www.pewclimate.org/global-warming-basics/.

30. Intergovernmental Panel on Climate Change, Working group 2. 2001. Summary for policymakers. *Climate change 2001: Impacts, adaptation, and vulnerability,* 6 http://www.grida.no/climate/ipcc_tar/wg2/pdf/wg2TARspm.pdf (last visited Oct. 27, 2003), (describing possible risks of climate change impacts). The possible impacts are climate-dependent, and the Working Group has not evaluated the full range of scenarios; Ibid. Climate change may be the most expensive disaster ever faced.; Simms, Andrew. 2002. *World disasters report 98.* Geneva: Int'l Federation of the Red Cross and Red Crescent.

31. Intergovernmental Panel on Climate Change, Working group 1. 2001. *Climate change 2001: The scientific basis,* sec.4.2. http://www.grida.no/climate/ipcc_tar/wg1/index.htm.

32. See generally Reilly, J. M., H.D. Jacoby and R.G. Prinn. 2003. Multi-gas contributors to global climate change: Climate impacts and mitigation costs of non-CO_2 gases. Pew Center of Global Climate Change, February. http://www.pewclimate.org/document.cfm?documentID=211 (last visited Sept. 4, 2004).

33. U.S. Department of Energy, EIA. 1999. *Emission of greenhouse gases in the United States, 1998.* Washington, DC: Government Printing Office.

34. Reitze, A. 2001. Global warming. *Environmental Law Reporter* 31: Washington DC, World Bank Group: 10253, 10255.

35. U.S. Department of Energy, EIA. 2004. *International Energy Outlook 2003.* http://www.eia.doe.gov/oiaf/ieo/.

36. Ibid.

37. Reitze, 2001 (pg. 10254). CO_2 levels have increased from 270–280 ppm in preindustrial times to more than 360 ppm in 1999. Nitrous oxide levels increased from 270 ppm to 310 ppm and methane concentrations have increased from 700 ppb to 1,700 ppb over the same period.

38. Ibid.

39. Ibid.

40. Ibid., 10255.

41. *World energy outlook 2002.* Nearly 25% of CO_2 emissions are from deforestation and natural processes.

42. *South China Morning Post.* 2003. One in four Beijingers now owns a car. August 6; *The Times of India.* 2003. LPG plan for rickshaws could clear air of toxins. August 23.

43. Uranium Institute. 1997. Responding to global climate change: The potential contribution of nuclear power. (position paper) http://www.world-nuclear.org/climate.htm (last visited Aug. 15, 2004).

44. French, H. W. 2004. China's boom brings fear of an electricity breakdown. *New York Times.* July 5.

45. Ibid.

46. U.S. Dept. of Energy. India country analysis brief. http://www.eia.doe.gov/cabs/india.html.

47. U.S. Dept. of Energy. Vietnam country analysis brief. http://www.eia.doe.gov/cabs/vietnam.html.

48. Intergovernmental Panel on Climate Change, Working group 1. 2001. Summary for policymakers. *Climate change 2001: The scientific basis*, 7. http://www.grida.no/climate/ipcc_tar/wg1/pdf/WG1_TAR-FRONT.PDF (last visited Oct. 26, 2003).

49. Intergovernmental Panel on Climate Change, WG1 Summary, 2.

50. Intergovernmental Panel on Climate Change, WG1 Summary, 14.

51. Intergovernmental Panel on Climate Change, WG2 Summary, 5.

52. Intergovernmental Panel on Climate Change, WG2 Summary, 10, 13 (assessing risks to human systems from flooding resulting from climate change); Ibid., 5. See also Kirshen, P., and R. Matthias. 2002. Dynamic investigation into climate change impacts on urban infrastructure 2. Paper presented at Western Regional Science Association annual meeting, Feb. 18–20, (discussing interrelationships between climate change and urban infrastructure and assessing possible impacts), http://www.puaf.umd.edu/faculty/papers/ruth/ClimateChangeConf.pdf (last visited Jan. 22, 2004).

53. Wolfensohn, J. (President, World Bank). 1997. Speech at United Nations General Assembly, New York. http://Inweb18.worldbank.org/ESSD/envext.nsf/ByDocName/ClimateChange.

54. Intergovernmental Panel on Climate Change, WG2 Summary, 10 (discussing climate change impacts on insurance and financial services industries).

55. Intergovernmental Panel on Climate Change. 2001. United Nations Environment Programme: Working Group 3 (WG3) (Mitigation), Summary for Policymakers 3, http://www.grida.no/climate/ipcc_tar/wg3/pdf/WG3_SPM.pdf (last visited Oct. 28, 2003).

56. See Grégoire, N. and I. Prigogine. 1989. *Exploring Complexity*, 36–40. Freeman & Co. (discussing nonlinear behavior of climate system and possible amplifying mechanisms).

57. Intergovernmental Panel on Climate Change, WG3 Summary, 35, 3.

58. Senator James M. Inhofe (chairman), floor statement on the science of climate change, on July 28. 2003, to the Committee on Environment and Public Works, http://epw.senate.gov/pressitem.cfm?party=rep&id=212247.

59. Murray, I. 2003. An improved climate. *Washington Times*, Dec. 26.

60. Grossman, D. 2001. Dissent in the maelstrom. *Scientific American*. (November), http://www.sciam.com/article.cfm?SID=mail&articleID=00095B0D-C331-1C6E-84A9809EC588EF21.

61. Appell, D. 2003. Hot words: A claim of nonhuman-induced global warming sparks debate. *Scientific American* (June 24), http://www.sciam.com/article.cfm?chanID=sa004&articleID=000829C7-70D9-1EF7-A6B8809EC588EEDF&pageNumber=1&catID=4.

62. Senator James M. Inhofe, 2003.

63. Ibid.

64. U.S. Council on Environmental Quality. 1981. *Global energy futures and the carbon dioxide problem*, viii. Washington, DC: Government Printing Office.

65. The U.S. voted 95–0 for the Byrd-Hagel Resolution, which opposed Kyoto and any climate change treaty that did not include binding GHG emission targets for developing nations on the same schedule as developed nations. *S. Rep. 105-54, 4 (1997)*.

66. *EEI Energy News*. September 26, 2002, 9.

67. Reitze, 2001 (pg. 10258).

68. U.S. Dept. of Energy, EIA. *Emissions of greenhouse gases in the United States 2002*. 2003. Washington, DC: Government Printing Office. This is because of increased deployment of renewable resources and cogeneration during this period and greater energy efficiency in U.S. manufacturing, production and delivery of services.

69. See World Bank Group. 2002. International environmental law: Concepts and issues. http://www4.worldbank.org/legal/legen/legen_iel.html.

70. International Energy Agency. 2004. *World energy outlook 2004*. Paris: International Energy Agency, http:// www.worldenergyoutlook.org.

71. Cabraal, A. 2004. *Towards a renewable energy future: A World Bank plan for action*. Washington DC, World Bank Group: 6.

72. See Nering, E. D. 2001. The mirage of a growing fuel supply. *New York Times*, June 4, A17.

73. Ibid.

74. Bruce, J. P. and E. F. Haites, eds.1996. *Climate change 1995: Economic and social dimensions of climate change*, 183. Cambridge: Cambridge University Press.

75. Sebenius, J. K. 1991. Designing negotiations toward a new regime: The case of global warming. *International Security* 15:110, 111. Climate change involves interplay between the planet's ecosystem and the human socioeconomic system, two vast and complex systems.

76. The precautionary principle also is articulated as "that if it is known that an action may cause profound and irreversible environmental damage, which permanently reduces the welfare of future generations, but the probability of such damage is not known, thin it is inequitable to act as if the probability is known." Charles Perrings, in Kysar, D. 2004. Climate change, cultural transformation, comprehensive rationality. *Boston College Environmental Affairs Law Review* 31:555, 565.

77. United Nations Conference on Environment and Development: Framework Convention on Climate Change, May 9, 1992, art. 3, S. Treaty Doc. No. 102-38, UN Doc. A/AC.237/18 (Part II) Add. 1, 81 I.L.M 849, http://unfccc.int/resource/docs/convkp/conveng.pdf (last visited Jan. 22, 2004) (emphasis added).

78. Stevenson, R. W. 2003. The struggle for Iraq: U.S. budget; 78% of Bush's postwar spending plan is military. *New York Times*, September 9, 12 (documenting $65.5 billion on military operations in Iraq).

79. *Economist*. 2003. The end of the oil age. October 25, 11, http://www.economist.com/printedition/index.cfm?d=20031025 (last visited 7/15/05)

80. U.S. Embassy Beijing. 2000. The costs of environmental degradation in China, http://www.usembassy-china.org.cn/sandt/CostofPollution-web.html (last visited Jan. 27, 2004); See also RAND Corporation. 2002. *Fault lines in China's economic terrain*, 109–16. (predicting increase in coal and natural gas use in event of oil price shock), http://www.rand.org/publications/MR/MR1686/ch6.pdf (last visited Sept. 21, 2003); see World Bank. 1997. *Clear water, blue skies: China's environment in the new century (analyzing prospects for environmental improvement and economic growth)*, http://www-wds.worldbank.org/servlet/WDSContentServer/WDSP/IB/1997/09/01/000009265_3980203115520/Rendered/PDF/multi_page.pdf (last visited Jan. 28, 2004).

81. U.S. Embassy Beijing, *Cost of environmental degradation*. Examples of China's environmental degradation include acid rain and rapidly deteriorating urban air quality. Nor are such problems confined to China alone: India's rapidly urbanizing society also faces the damage of acid rain.

82. *Economist*, 2003 (pg. 11).

83. International Energy Agency. 2004. *World energy outlook 2004*. Paris: International Energy Agency, http://www.worldenergyoutlook.org.

84. Ibid.

85. Cooper, D. E. 1999. The Kyoto Protocol and China: Global warming's sleeping giant. *Georgetown International Environmental. Law Review* 11: 401, 405.

86. International Energy Agency. 2004. *World energy outlook* 2004. Paris: International Energy Agency, http:// www.worldenergyoutlook.org.

87. Ferrey, S. *Law of Independent Power*, chap. 3, sec. 20.1.

88. Yardley, J. 2004. The world China's economic engine needs power (lots of it). *New York Times*, March 14, Week in review 3.

89. Boon-Siew, Y. and R. Rajaraman. 2004. Electricity in China: The latest reforms, *Electricity Journal*, (April). Per capita installed generation of 0.25 kW with per capita annual consumption of 1,078 kWh; Ibid. These are half the value of world averages and only approximately 10%–20% of average consumption in developed countries; Ibid.

90. Ibid.

91. Ibid.

92. International Conference for Renewable. 2004. International Action Programme. http://www.renewables2004.de/pdf/International_Action_Programme.pdf.

93. International Energy Agency. *World energy investment outlook 2003*, 2003.

94. Sim, S. 1996. Overtaking the West: Asia's teeming Urbanites. *The Straits Times* (Singapore), Dec. 9, 41.

95. U.S. Dept. of Energy, *India country analysis brief.*

96. CNN. 2002. Consumers underpin India's growth, Dec. 31, http://www.cnn.com/2002/BUSINESS/asia/12/31/india.gdp/.

97. World Bank. *Indonesia country report*, Apr. 20, 2004, http://siteresources.worldbank.org/INTINDONESIA/Resources/Country-Update/april2004.pdf.

98. French, 2004 (pg. 4).

99. International Energy Agency. *World energy investment outlook 2003*, 2003.

100. Fritsche and Matthes, Changing Course, 28, utilizing IEA data from World Energy Outlook 2000.

101. Ibid., utilizing IEA 2002 data.

102. For more on this topic, see Ferrey, *The Law of Independent Power*, chap. 5.

103. Fritsche and Matthes, 2003. (pg. 32, fig. 8) utilizing IEA data from 2002.

104. Ferrey, *The Law of Independent Power*, ch. 6:136.

105. Ibid., sec. 2:11.

106. The World Bank Group, *Renewable Energy for Development*, 2.

2. Renewable Energy as the Global Warming Solution

> Energy is the single most important problem facing humanity today. We must find an alternative to oil...the cheaper, cleaner, and more universally available this new energy technology is, the better we will be able to avoid human suffering, and the major upheavals of war and terrorism.
>
> — Richard Smalley, Nobel Laureate, 2002[1]

Renewable Energy Options and Alternatives

The renewable source

The technologies of choice are emissions-free, renewable energy. Solar energy is the source of all energy on earth; it creates wind and water movement and ultimately creating plants,[2] biomass, and animals that become fossil fuels when their organic matter decays. Our sun is one of a family of stars known to astronomers as *yellow dwarfs*. The sun is a G2 dwarf of unremarkable stellar magnitude. It is powered by several kinds of fusion reactions that have consumed 11 billion pounds of hydrogen each second for the past 4 to 5 billion years and is expected to continue for another 4 to 5 billion years. At its core the sun consumes 4.3 million tons of matter every second, releasing 3.89 times 10^{26} joules of nuclear energy. Although the energy output of the sun in the direction of the earth is approximately 1,300 W/m^2 at its source, one-third is reflected back into space by the earth's atmosphere, yielding as much as 1,000 W/m^2 at the surface of the earth at noon on a cloudless day; on average over the course of a year, around 170 W/m² of solar radiation reaches the earth's oceans, and roughly 180 W/m² reaches the land surfaces.[3]

Human capture of this energy is neither efficient nor prodigious. Energy used by humankind on the earth equals only approximately 0.01% of the total solar energy reaching the earth. In fact, no nation on earth uses more energy than the energy content contained in the sunlight that strikes its existing buildings every day. The solar energy that falls on roads in the United States each year contains roughly as much energy content as all the fossil fuel consumed in the world during that same year.

Unlike finite fossil fuels, solar energy represents a constantly replenished flow, rather than an existing stock that is diminished by its use. Tomorrow, the earth will have exactly as much solar energy as it has today, regardless of how much solar energy is used and consumed each day. By contrast, burning a barrel of oil or a cubic meter of natural gas diminishes permanently that quantity of fossil fuels for the next day and for future generations.

In 1997 all nations on earth consumed 26.4 billion barrels of oil, 81.7 trillion cubic feet of natural gas, and 5.2 billion tons of coal—all of which are decayed organic matter previously brought to life by the sun. Although many nations, particularly developing nations, have no significant reserves of oil, coal, or natural gas, every nation has solar energy in some form—sunlight, wind, ocean wave power, and so forth. Every nation has some indigenous renewable energy resource, allowing for energy independence and providing a source for domestic economic development. Although the commercial and national interests involved in fossil fuel are extremely concentrated, solar energy interests and flows are much more decentralized and diverse.

Solar radiation creates winds. It is converted to heat and generates pressure differences required to maintain motion in the atmosphere. In addition to solar radiation, a second source of heat for the earth is the slow release of the earth's basal heat in the form of the cooling of the earth's core. Geothermal energy sources tap this basal heat, and sudden volcanic eruptions violently release this heat.[4]

In contrast to fossil fuel sources, solar, wind, and geothermal energy produce no emissions during power generation. Their use can offset emissions of CO_2 and other criteria pollutants that fossil fuel-fired plants would otherwise generate. Still, solar and wind energy are intermittent resources and as such cannot be used reliably to meet base load electricity requirements without integration with electric energy storage. More traditional renewable generation resources such as hydroelectric or geothermal power generation and biomass fuels are round-the-clock transmittable resources that can supply base load power resources.

Only location limits solar, wind, and geothermal resources. Although fossil fuel-fired plants can be sited anywhere with appropriate fuel delivery and electricity transmission infrastructure, large renewable power plants can only be sited where renewable sources are present in large enough amounts and concentrations to make the capital investment in generation facilities feasible. But unlike fossil fuel-fired generation facilities, renewable energy resources are not limited by a finite fuel supply.

Renewable projects can either be implemented as on-grid (transmission) projects or off-grid projects. Unless elaborate controls are placed on the system, stand-alone dispersed off-grid systems can either provide insufficient power at peak demand or excess during off-peak hours.[5] Off-grid renewable energy project revenues are often less secure or unsecured, increasing financing rates.[6]

Renewable energy resources are broadly available across Asia and in many developing nations.[7] China committed in 2004 to generate 10% of its total power needs from renewable resources by 2010 as part of voluntary targets that will promote primarily hydropower.[8]

The Renewable Technologies for Developing Nations

Hydroelectric power

Hydroelectric power plants convert the kinetic energy contained in falling water into electricity. Hydro power is currently the world's largest renewable source of electricity, accounting for 6% of worldwide energy supply or approximately 15% of the world's electricity. By the middle of the 20th century, 5,000 large dams had been constructed across the world, with 75% of these in industrialized countries; by the end of the 20th century, there were more than 45,000 large dams in more than 140 countries.[9] Hydro power supplies more than 90% of total national electric central supply in two dozen countries and more than half of centralized electric supply in 63 countries.[10]

The theoretical size of worldwide hydro power capacity is approximately four times greater than that which has been exploited at this time. Much of the remaining hydro potential in the world exists in the developing countries of Africa and Asia. This potential can be tapped either in larger dam projects or in a series of smaller hydroelectric projects at village scale. This modular flexibility is one of the signatures of renewable resources. Traditionally regarded as a cheap and clean source of electricity, many large hydroelectric schemes being planned today encounter a great deal of opposition from environmental groups and native people. Planning the scale and location of exploitation of a river system is a key issue.

Falling water turns a turbine that converts the water's energy into mechanical power. A generator converts the rotation of the water turbines into electricity. The amount of electricity that can be generated at a hydroelectric plant depends on the vertical distance the water falls, called the *head*, and the *flow rate*, measured as volume-per-unit time. The electricity produced is proportional to the product of the head and the rate of flow: POWER (kW) = 5.9 × FLOW × HEAD[11]

The most common design for hydro plants uses a dam to store water at elevation. A dam's ability to store water during rainy periods and release it during dry periods results in reliable electricity production.[12] Pumped storage is another form of hydroelectric power. Pumped storage facilities use excess electrical system capacity, generally available at night, to pump water from one reservoir to another reservoir at a higher elevation. During periods of peak electrical demand, water from the higher reservoir is released through turbines to the lower reservoir and electricity is produced.[13]

Hydroelectric power plants do not emit any of the standard atmospheric pollutants such as carbon dioxide or sulfur dioxide given off by fossil fuel-fired power plants. However, the environmental impacts of large hydro projects have been called into question.[14] The greatest and most obvious impact of large hydroelectric dams is their vast flood areas, which can threaten ecosystems. Large dams and reservoirs can have other impacts on a watershed[15] (see table 2–1).

Table 2–1: Summary of Environmental Costs for Various Renewable Energy Technologies

Technology Type	Cents/kWh
Solar	0 to 0.4
Wind	0 to 0.1
Biomass	0 to 0.7

Solar photovoltaic energy

Photovoltaic (PV) materials use the sun's energy to produce electricity and therefore result in none of the greenhouse or acid gas emissions associated with electricity generated by the combustion of fossil fuels.[16] The amount of solar energy reaching the earth each year is many times greater than worldwide energy demand, although it varies with location, time of day, and the season. Sunlight is also a widely dispersed resource, and photovoltaics can capture energy from the sun virtually anywhere on earth.

The basic PV building block is the photovoltaic cell. It is referred to as a *cell*, because it produces direct current (DC) electricity like a battery, converting energy from the sun directly into electricity. In practical applications of photovoltaics, groups of cells are joined together to form a module, and modules may be connected into an array. These cells, modules, and arrays can provide electricity in any quantity, ranging from a few milliwatts powering a calculator to several megawatts, the size of a large power plant.

Using PV to supply electricity to electrical transmission grids can be accomplished either through vast arrays of PV modules in a centralized location acting in the same manner as a traditional power plant, or by decentralized arrays of photovoltaics on the roofs of houses and buildings. Electricity produced by photovoltaics for electrical grids currently costs roughly 30 cents per kWh. Electricity from traditional coal-fired power plants costs approximately 6 cents per kWh. The costs of PV cells will have to come down even further before photovoltaics are widely connected to electrical grids.

Sunlight is the fuel for photovoltaics, so the cost of manufacturing the cells is the main cost of producing electricity. Electricity storage, or using electricity immediately when it is generated, is necessary for PV cell applications because they generate electricity only when the sun is shining. The original and still the most common semiconducting material used in PV cells is single crystal silicon.[17]

Electricity produced from photovoltaics has a far smaller impact on the environment than traditional methods of electric generation. During their operation, PV cells use no fuel other than sunlight, give off no atmospheric or water pollutants, and require no cooling water. Unlike fossil-fired power plants, photovoltaics do not contribute to global warming or acid rain. Similarly photovoltaics present no radioactive risks.

The only negative environmental impacts associated with photovoltaics are potentially toxic chemicals used during their manufacture, and the amount of land used to produce electricity, which is minimized if PV arrays are placed on buildings, as is common in many applications.[18] Industrial-scale PV arrays do use a large amount of land, but the amount is comparable to other energy production technologies such as coal or nuclear when fuel production and storage are taken into account. Photovoltaic solar energy systems exhibit environmental[19] externalities that range from 0 to 0.4 cents/kWh, as illustrated in table 2–1. Photovoltaic is ideal for remote applications and therefore do not require as much transmission capacity as traditional centralized power supply.

A related technology, the solar-thermal energy facility, collects solar energy and converts it into electricity. The plants' production process happens in two stages. Initially, the sun's energy is converted into high-temperature heat using various mirror configurations. Then the heat is used to boil water to run a conventional steam turbine.[20] Where it shines, the sun's power is abundant. In the United States for example, parabolic trough systems covering approximately 9% of Nevada—a plot of land approximately 1,000 square miles—could generate enough electrical power for the entire country.[21]

Wind power

Wind power uses the spinning turbines of industrial-scale windmills to convert wind into electricity.[22] Wind turbines capture the wind's energy with two or three propeller-like blades, which are mounted on a rotor to generate electricity. The turbines sit high atop towers, taking advantage of the stronger and less turbulent wind at 100 feet (30 meters) or more above ground. Modern large wind turbines generally have three blades and sit atop towers ranging from 150 feet tall for onshore turbines to 300 feet tall for offshore turbines. These high turbines are designed to take advantage of stronger winds above ground level.

A blade acts much like an airplane wing. When the wind blows, a pocket of low-pressure air forms on the downwind side of the blade and pulls the blade toward it, causing the rotor to turn; this is called *lift*. The force of the lift is actually much stronger than the wind's force against the front side of the blade, which is called *drag*. The combination of lift and drag causes the rotor to spin like a propeller, and the turning shaft spins a generator to make electricity.

The amount of energy that can be captured by a wind turbine increases as the cube of wind velocity. For example, the potential amount of energy, which could be captured by a wind turbine in an area where wind speeds average 20 km/h is approximately 2.5 times that in a region where the average wind speed is 15 km/h. The more electricity a single wind turbine can generate in a year, the more economical it is to operate.

Wind machines can be sited in a variety of sizes. Modern wind turbines come in a wide range of sizes, from small 100-watt units designed to provide power for single homes or cottages, to huge turbines with blade diameters more than 50 meters, generating more than 3 MW of electricity. Wind farms must be set up strategically to avoid *wake* losses, that is, turbulence losses because of the turbines' interference with each other's wind.

The *fuel* for wind turbines is free. The current cost of wind generated electricity is approximately 5 to 9 cents per kWh. This is similar to the costs of generating electricity from fossil fuels and is cheaper than the cost of electricity from most recent nuclear power plants. Unlike in a fossil plant, however, the power is not transmittable when it is needed because the power is generated when the wind blows. Although there have been some technological challenges to wind power, especially with regards to the robustness of the turbine technology and the intermittent nature of the wind resource, the cents-per-kilowatt cost of wind energy production has fallen dramatically in recent years as countries around the world have added wind power to their energy portfolios.

Wind energy facilities create noise and land-use externalities. There has been concern with bird kills resulting from wind farms. The cost of these externalities are approximately 0.1 cents/kWh, as illustrated in table 2-1. Globally, installed wind generation capacity has increased by an average 25% annually since 1990.

Biomass energy

Before the industrial revolution, biomass—all land- and water-based vegetation and all organic wastes—fulfilled almost all of humankind's energy needs. Since the industrial revolution, the majority of the developed world's energy requirements have been met by the combustion of fossil fuels. Biomass, however, is still the predominant form of energy used by people in developing countries.

The three basic sources of biomass energy are municipal and industrial wastes, agricultural crop residues, and energy plantations. Crops that have been specifically grown for energy include sugar cane, corn, sugar beets, grains, elephant grass, and kelp (seaweed), to mention a few.[23] Virtually all crops leave some form of organic residues after their primary use. These organic residues, as well as animal excrement, can be used for energy production through direct combustion or biochemical conversion. Crop residues are usually returned to the soil, and the resulting compost helps maintain soil nutrients, soil porosity, water infiltration and storage, as well as reducing soil erosion. An increased scale of use for fuel may have significant environmental impacts, the most serious being those of lost soil fertility and soil erosion.

Direct combustion is the simplest, cheapest, and most common method of obtaining energy from biomass. Combustion heat can provide space or process heat, water heating, or through the use of a steam turbine, electricity. In the developing world, many types of biomass, such as dung and agricultural wastes, are burned for cooking and heating.

Biomass products can be converted into commercial fuels that substitute for fossil fuels. Both biochemical and thermal conversions to produce gaseous, liquid, and solid fuels, which have high energy content, are easily transportable, and are therefore suitable for use as commercial fuels. Alcoholic fermentation of biomass produces liquid fuels and *anaerobic* digestion or fermentation results in biogas. For example, fermenting sugarcane and corn to produce ethanol for use in internal combustion engines has been practiced for years.

The organic fraction of almost any form of biomass, including sewage sludge, animal wastes, and industrial effluents, can be broken down through anaerobic digestion into CH_4 and carbon dioxide. This *biogas* is a reasonably clean-burning fuel, which can be captured and put to many different end uses such as cooking, heating, or electrical generation.

The consensus is that biomass fuels used in a sustainable manner result in no net increase in atmospheric CO_2. This is based on the assumption that all the CO_2 given off by the use of biomass fuels for energy was recently taken in from the atmosphere by photosynthesis. Increased substitution of fossil fuels with biomass based fuels would therefore help reduce the potential for global warming, caused by increased atmospheric concentrations of CO_2.

Biomass energy facilities, depending on their fuel sources, emit a variety of criteria air pollutants from their combustion of organic materials. Replacing fossil fuels with biomass can reduce the potential for acid rain; moreover, the lower burn temperatures of biomass produce less nitrogen oxide (NO_x) per BTU than fossil fuels. Biomass generally contains less than 0.1% sulfur by weight, whereas low sulfur coal has 0.5% to 4% sulfur. The environmental externalities of biomass power facilities range from 0 to 0.7 cents/kWh, as illustrated in table 2–1.

Landfill gas

Landfill gas is produced naturally by anaerobic decomposition of organic matter in landfills. It is composed of 40% to 60% methane (CH_4), with the remainder principally CO_2. Each metric ton of solid waste produces approximately 70 cubic meters of landfill gas. Landfill gas has a gross heating value of approximately half of natural gas.

Methods of collecting landfill gas vary with the design of a particular landfill. Methane is potentially explosive, so many modern landfills have been built with CH_4 collection systems to flare the collected gas. Landfill gas can be used to produce electric energy through a reciprocating engine or gas turbine. Burning landfill gas produces the same environmental emissions as burning natural gas, including NO_x and CO_2 emissions.

Geothermal

Generators have been using geothermal steam reservoirs to make electricity in California and Italy for more than 30 years. It is also exploited in the Philippines, Indonesia, Japan, and other Asian Pacific Rim countries. The range of viable sites is expanded with a development-stage technology involving deep drilling to where magma from the earth's mantle comes unusually close to the surface, pumping water down the hole and using the resulting steam to power a turbine.[24] Geothermal energy sources have minimal environmental impacts, which while limited, include air pollution and noise when the effluents are pumped back into the reservoirs, as is the normal practice.[25]

Ocean energy

Ocean energy, still far from reaching the commercial stage, draws on the mechanical energy of ocean waves, tides, or thermal energy stored in the ocean. Each day the oceans absorb enough heat from the sun to equal the thermal energy contained in 250 billion barrels of oil.[26] The sun warms the surface water much more than the deep ocean water, and this temperature difference stores thermal energy. Ocean thermal energy conversion (OTEC) systems convert this thermal energy into electricity, often while producing desalinated water.

The three types of electricity conversion systems are closed cycle, open cycle, and hybrid. Closed-cycle systems use the ocean's warm surface water to vaporize a working fluid with a low-boiling point, such as ammonia. The vapor expands and turns a turbine. The turbine then activates a generator to produce electricity. Open-cycle systems actually boil the seawater by operating at low pressures. This produces steam that passes through a turbine/generator. Hybrid systems combine both closed-cycle and open-cycle systems.

Ocean mechanical energy is quite different from ocean thermal energy. Although the sun affects all ocean activity, tides are driven primarily by the gravitational pull of the moon, and waves are driven primarily by the winds. The estimated power of waves breaking on the world's coastlines is 2 to 3 million MW.[27] In some locations, wave energy density can average 65 MW per mile of coastline.[28] The potential for deriving energy from the ocean tides is also great. France's La Rance station is the largest facility harnessing the power of tidal basins, generating 240 megawatts of power.[29]

A barrage (dam) is typically used to convert tidal energy into electricity by forcing the water through turbines to activate a generator. Tidal energy is derived from the difference in high and two low tides. Coastal areas consistently experience two tidal cycles over a period of slightly more than 24 hours, but to effectively convert tidal energy into electricity with existing technology, the difference between high and low tides must be at least five meters, or more than 16 feet.[30] There are only approximately 40 sites on the planet with tidal ranges of this magnitude.

For wave energy conversion, there are three basic systems. They are channel systems that funnel the waves into reservoirs, float systems that drive hydraulic pumps, and oscillating water column systems that use the waves to compress air within a container. The mechanical power created from these systems either directly activates a generator or transfers to a working fluid, water or air, which then drives a turbine/generator.

Hydrogen fuel cells

Hydrogen is the third most abundant element on the earth's surface, where it is found primarily in water and organic compounds. It is also the most plentiful element in the universe. Hydrogen is the simplest element; an atom that consists of only one proton and one electron. Despite its simplicity and abundance, hydrogen doesn't occur naturally as a gas on the earth; it is always combined with other elements. In addition to water, hydrogen is also found in many organic compounds, notably the *hydrocarbons* that make up many of our fuels, such as gasoline, natural gas, methanol, and propane.

Hydrogen can be made by separating it from hydrocarbons by applying heat, a process known as *reforming* hydrogen. Currently, most hydrogen is made this way from natural gas. An electrical current can also be used to separate water into its components of oxygen and hydrogen. Some algae and bacteria, using sunlight as their energy source, even give off hydrogen under certain conditions.

Hydrogen is high in energy, yet an engine that burns pure hydrogen produces almost no pollution. NASA has used liquid hydrogen since the 1970s to propel the space shuttle and other rockets into orbit. Hydrogen fuel cells power the shuttle's electrical systems and produce water as a byproduct, which the crew drinks.

In the future, hydrogen could also join electricity as an important energy carrier. An energy carrier stores, moves, and delivers energy in a usable form to consumers. Many renewable energy sources, like the sun or wind, are not available to produce energy all the time, but hydrogen can store this energy until it is needed and it can be transported to where it is needed by underground pipelines.

Installed Capacity of Renewable Energy for Electricity Production

Total installed capacity of renewable energy, excluding large hydropower, for electricity generation was 142 GW worldwide, of which 58 GW was in developing countries (see table 2–2). Although this installed capacity is but a tiny fraction of the 3,700 MW of total electricity generation capacity, installations of some technologies such as wind and solar photovoltaics are growing at more than 25% per year, albeit from a small base.

Comparative Environmental Emissions

Renewable energy alternatives

Many renewable energy projects, other than those using biomass fuels, do not use combustion of fuels to produce electricity. They create mechanical shaft power from the movement of wind or water, tap naturally produced geologic steam, or employ solar energy to induce direct current on a chemical surface. Because renewable energy alternatives—solar, wind, hydro, geothermal—do not involve combustion to produce electric energy, they do not emit these various pollutants or GHGs during their operation. Where biomass, waste, or landfill gas is burned, there is combustion.

Table 2–2: Grid-based renewable power capacity as of 2003

Generation type (GW)	Capacity in all countries (GW)	Capacity in developing countries[a]
Small hydro power[b]	56	33
Wind power	40	3
Biomass power[c]	35	18
Geothermal power	9	4
Solar photovoltaic power (grid-connected)	1.1	< 0.1
Solar thermal power	0.4	0
Total renewable power capacity	142	58
For comparison – large hydro power[d] – total electric power capacity	730 3700	340 1300

a Developing countries are non-OECD countries plus Mexico, South Korea and Turkey, excluding countries with economies in transition. Martinot et al. (2002) included economies in transition in these totals, reflecting all countries eligible for World Bank development assistance.

b Definitions of small hydro vary by country. They usually cover hydro up to 10 MW, although this figure is up to 25 MW in India and up to 30 MW in China—thus global totals can differ greatly depending on what is counted.

c Biomass power figures exclude municipal solid waste combustion and landfill gas.

d Published hydro power figures assumed to include both large and small hydro, except in China, where these are reported separately. Total hydro is the sum of small and large hydro.

Source: http://www.jxj.com/magsandj/rew/2004_05/indicators_of_investment.html

In most applications, distributed electric production tends to decrease air emissions.[31] Renewable technologies are cleaner than fossil fuel–fired technologies. Microturbines are cleaner than diesel engines. Neither microturbines nor diesel engines are as clean as the new state-of-the art gas-fired combined cycle generation technologies. To the degree that distributed generators displace older coal-fired or other older and less efficient generation, however, they create net environmental advantages.[32]

Notably, many of these renewable combustion sources can be environmentally preferable to burning fossil fuels. First, these sources are renewable, unlike fossil fuels, and producing electricity with them is the best alternative for their necessary disposal. In most developing countries biomass or agricultural waste, such as from sugar production from cane, is burned when the process is complete. GHGs and other pollutants will be released to the atmosphere in any event. If this burning is shifted to a boiler from which electricity can be produced, at least there is an electricity gain from that necessary combustion of waste.

Landfill gas is composed approximately 50% each of CH_4 and CO_2. Both are potent GHGs. If left in the landfill, these gases will eventually escape to the atmosphere. Otherwise, if extracted and used to fuel electricity production, the CH_4 burned is converted to CO_2. This is preferable in at least three regards. First, CO_2 is a much less potent GHG than CH_4, so this conversion is a net environmental gain in terms of global warming concerns. Second, electricity is produced that offsets the potential need to produce it by burning increasingly taxed fossil fuel supplies. Third, if left in landfills, landfill gas can migrate into nearby buildings causing the risk of explosion. And if used in a cogeneration system, where after producing electricity thermal energy is captured for use, combustion of these waste fuels is more efficient.

With renewable resources, putting aside the currently greater cost associated with their use, the environmental issue typically is one of land use. To produce a large amount of power from wind or solar energy, a significant area must be devoted to installation of equipment. The area covered by wind turbines or solar collectors is greater than the area covered by a fossil fuel-fired power plant that would produce an equivalent amount of power.

Conventional power technologies

Nevertheless we do combust prodigious quantities of fossil fuels. In fact, fire, or combustion, is a relatively inefficient and environmentally polluting means to exploit the earth's potential electric energy sources. Approximately three-quarters of the anthropogenic sources of carbon in the atmosphere is the result of the combustion of fossil fuels, whereas 25% is the result of deforestation and resultant inability of the biosphere to assimilate and reprocess this chemical compound.[33] The electric power sector is projected to account for 35% of anthropogenic CO_2 emissions.[34] Conventional production of centralized fossil fuel–fired electricity by electric utilities is responsible for 68% of sulfur dioxide (SO_2) emissions and 33% of NO_x emissions.[35]

Environmental costs associated with power plants occur at each of three stages of energy extraction, use of energy sources,[36] and back-end residual costs.[37] The primary impacts on human health from direct production of electric energy are from emissions of CO_2 and the pollutants SO_2, NO_x, O_3, volatile organic compounds, particulates, and acid deposition. Although these pollutants are not as directly toxic to the environment as mercury or lead, for example, they do cause a wide range of serious environmental impacts that have a high economic cost in terms of both natural resources and human health, including acid rain, smog, and respiratory illness.

Although these pollutants were initially thought to impact principally the environment near their source of emission (unlike GHGs), the most recent scientific research suggests that these pollutants disperse globally. Emissions from the United States drift to Europe, Asian emissions drift to the United States, and so forth. Conventional power facilities can exert environmental impacts on health and the environment in the form of water pollution[38] and impairment of land uses.[39] Briefly summarized below, each of these by-products of combusting fossil fuels to generate electricity have significant impacts on the environment and human health.

NO_x. Nitrogen oxide forms during the combustion of fossil fuel, resulting from the conversion of chemically bound nitrogen in certain fuels, such as coal or oil, and by the thermal fixation of atmospheric nitrogen during fossil fuel combustion. Nitrogen oxide causes a wide variety of health and environmental impacts because of various compounds and derivatives in the family of nitrogen oxides, including nitrogen dioxide, nitric acid, nitrous oxide, nitrates, and nitric oxide. Nitrogen oxide reacts with ammonia, moisture, and other compounds to form nitric acid and related particles.

Human health concerns include effects on breathing and the respiratory system, damage to lung tissue, and premature death. Small particles penetrate deeply into sensitive parts of the lungs and can cause or worsen respiratory disease, such as emphysema and bronchitis, and aggravate existing heart disease.

Increased nitrogen loading in water bodies, particularly coastal estuaries, upsets the chemical balance of nutrients used by aquatic plants and animals. Additional nitrogen accelerates *eutrophication*, which leads to oxygen depletion and reduces fish and shellfish populations. In the air, NO_x reacts readily with common organic chemicals and even O_3 to form a wide variety of toxic products, some of which may cause biological mutations. Examples of these chemicals include the nitrate radical, nitroarenes, and nitrosamines.

SO_2. SO_2 is formed by the oxidation of sulfur in fuel, and the quantity released is a direct function of the sulfur content of any given fuel. More than 65% of SO_2 released to the air, or more than 13 million tons per year, comes from electric utilities, especially those that burn coal. China's coal plants burn cheap coal with high sulfur content. The resulting SO_2 emissions significantly deteriorate air quality and acid rain. Indeed, SO_2 has emerged as the greatest immediate environmental threat to China's urban areas since leaded gasoline. Acid rain damage in China because of use of high sulfur coal was estimated at more or less 2% of GDP (more than $13 billion in 1995). Recognizing these costs, use of lower sulfur coals is now mandated.[40]

More generally, high levels of SO_2 in the air can cause temporary breathing difficulty for people with asthma who are active outdoors. Longer-term exposures to high levels of SO_2 gas and particles cause respiratory illness and aggravate existing heart disease. It also reacts with other chemicals in the air to form tiny sulfate particles. When inhaled they collect in the lungs and are associated with increased respiratory symptoms and disease, difficulty in breathing, and premature death. Sulfur dioxide and nitrogen oxides react with other substances in the air to form acids, which fall to earth as rain, fog, snow, or dry particles.

VOCs. Volatile organic compounds (VOCs), also sometimes referred to as hydrocarbons or non-methane organic gases (NMOGs), are part of emissions from coal and oil combustion. Their ability to cause health effects varies greatly from those that are highly toxic to those with no known health effect. As with other pollutants, the extent and nature of the health effect will depend on many factors, including the level of exposure and length of time exposed. Eye and respiratory tract irritation, headaches, dizziness, visual disorders, and memory impairment are among the immediate symptoms that some people have experienced soon after exposure to various organic compounds. Many organic compounds are known to cause cancer in animals; some are suspected of causing, or are known to cause, cancer in humans.

VOCs are also an important component of smog. Smog (ground level O_3) is formed when VOCs and NO_x react in the presence of heat and sunlight. Children, people with lung diseases such as asthma, and people who work or exercise outside are susceptible to adverse effects such as damage to lung tissue and reduction in lung function. Ozone can be transported by wind currents and cause health impacts far from original sources. Other impacts from ozone include damaged vegetation and reduced crop yields.

CO. Carbon monoxide (CO) is an inert gas formed by combustion when there is insufficient fuel, after residence time in the combustion chamber of a power plant at high temperatures, when the combustion gas temperature is too low, or if there is insufficient oxygen to complete oxidation of a hydrocarbon molecule. Carbon monoxide is not known to have the most serious environmental impacts of the other pollutants discussed here, but in concentration it can have a severe impact on human health. Carbon monoxide enters the bloodstream through the lungs and reduces oxygen delivery to the body's organs and tissues because it replaces the oxygen atom in the center of the hemoglobin molecule.

The health threat from levels of CO sometimes found in the ambient air is most serious for those who suffer from cardiovascular disease. At much higher levels of exposure not commonly found in surrounding air, CO can be poisonous, and even healthy individuals may be affected.

Visual impairment, reduced work capacity, reduced manual dexterity, poor learning ability, and difficulty in performing complex tasks are all associated with exposure to elevated CO levels.

Acidification

Some of these pollutants react together to cause significant impacts on natural systems and human health. Acid rain is an excellent example with unique impacts. NO_x and SO_2 react with VOCs in the air to form acid rain (or acid snow or fog). Acid rain may be carried by wind for hundreds of miles and has a wide variety of environmental and health impacts. Soils, surface waters such as lakes and rivers, and forests can all be damaged by acid rain.

The actual interactions between aquatic organisms (such as fish, crustaceans, insects, and amphibians) and changes in water chemistry are extremely complex. Acidification can hinder the ability of aquatic organisms to reproduce. This is especially true for fish and amphibians that spawn in streams or shallow bays in the early spring. This *acid shock* can kill the eggs of many species. In addition to reproductive failures, acidification can reduce the amount of calcium available to vertebrates such as fish, as well as increasing the concentration of toxic heavy metals in surface waters. Acidification can eventually result in a *dead* lake.

The effects of acid rain on terrestrial plants and animals are also both complex and potentially devastating. Soils can be damaged by removing needed nutrients and dissolving toxic heavy metals from the soil. Metals such as aluminum can get into the roots of plants and prevent the uptake of other crucial nutrients. Forests in areas with sensitive soils can be severely affected by acidification. Acid rain can damage the foliage of trees and retard leaf growth.

Acid rain can also cause human health concerns and damage to buildings. Acid rain can aggravate respiratory ailments, such as bronchitis and asthma. Humans may also be affected by drinking water that contains higher levels of toxic metals which have been dissolved from soils and pipes by the increased acidity of drinking water supplies. Construction materials such as limestone, marble, and sandstone can also be damaged by acid rain, resulting in eroded buildings and monuments.

Fuel Accidents

Also worthy of mention are oil spills, a form of pollution that, while not directly related to fossil fuel emissions, are directly related to fossil fuel use. Oil spills have a devastating impact on marine organisms, and oil from open ocean spills can end up contaminating beaches hundreds of miles away. Oil spills can harm marine life by poisoning after ingestion, direct contact, and destroying habitats.

The negative effects of ingesting toxic levels of oil are poorly understood for many specific organisms, especially microorganisms such as plankton, bottom-dwelling organisms, and larval fish. The effects on larger creatures such as fish and marine mammals are much more fully documented. Fish ingest large amounts of oil through their gills. If this does not kill them directly, it can inhibit their ability to reproduce or result in deformed offspring. Especially vulnerable are slow-moving shellfish such as clams, oysters, and mussels.

Marine mammals and birds in direct contact with oil slicks often ingest a great deal of oil while attempting to clean themselves. Carnivorous animals and birds that eat the carcasses of other oiled creatures also end up ingesting potentially toxic amounts of oil. Ingesting oil can destroy an animal's internal organs and interferes with the reproductive process. The famous Exxon Valdez oil spill in Alaska's Prince William Sound resulted in the deaths of 15,000 otters, predominantly as a result of ingesting oil.

Birds and marine mammals can also be killed by direct exposure to oil. Oil can clog a bird's feathers making it impossible for the bird to fly and so heavy they may simply sink rather than float. Oil also prevents a bird's feathers from keeping it warm. In colder climates, countless oiled birds die of hypothermia. Similarly, mammals in cold waters can also die of hypothermia as their fur loses its insulating ability once it has been covered in oil. Finally, when tar-like clumps of oil sink to the bottom, living conditions for benthic organisms are destroyed, as are spawning sites for many types of fish and shellfish.

Decentralization of Electric Generation

Distributed generation, where electricity is produced and consumed on-site off the grid in a decentralized manner, represents a paradigm shift. Distributed electric generation resources include the renewable resources listed previously, as well as cogeneration resources that may be fired by fossil or renewable fuels. Cogeneration facilities should cause fewer environmental impacts than equivalent megawatts of conventional power production.

This savings occurs because cogeneration facilities simultaneously produce electricity and usable thermal energy by the same continuous process, recapturing and using more efficiently energy that would otherwise be wasted. This substitution of an integrated cogeneration technology, in lieu of conventional separate electricity and thermal energy production technologies, should save 15% to 25% of the energy input otherwise consumed by separate energy production configurations.[41] This can also be true of small distributed generator energy production which, although they do not realize the efficiency inherent in cogeneration technologies, combust waste or alternative energy.

The efficiency of distributed generation improves greatly when the thermal by-product of the combustion process is productively used for thermal application. With very efficient cogeneration technology, the CO_2 emissions can actually be less than for large central station combined-cycle generation and NO_x emissions can be comparable.[42] Various cogeneration technologies can reduce the levels of sulfur oxides (such as SO_2),[43] particulate matter[44], carbon dioxide, and NO_x per unit of useful energy output, although certain technology configurations can also increase the discharge of these critical emissions.[45] Typical air emission influences of technologies, without added emission controls, are displayed in table 2–3.

Table 2–3: Air Quality Impacts of Cogeneration

Technological Characteristics	Direct Physical Effects	Impact on Air Quality Positive, Negative or Mixed
Increased Efficiency	Reduction in total emissions per unit of energy produced	Positive
Smaller Scale of QF deployment	Change in emissions levels Change in level of environmental control, usually less Lesser emissions Stack height	Mixed or negative Negative Mixed or negative
Change in energy production technology	Change in emissions and type of pollutants	Mixed
Change of fuels	Change in emissions and type of pollutants	Mixed or positive
Change of location of electricity generation	Change in location of emissions, density and distribution	Mixed

Distributed Generation Reliability

In 2003 the Congressional Budget Office concluded, "Distributed generation, the small-scale production of electricity at or near customers' homes and businesses, has the potential to improve the reliability of the power supply, reduce the cost of electricity, and lower emissions of air pollutants."[46] Distributed resources have the potential to increase reliability of a national or regional electric system because distributed generation occurs throughout the electric system; distributed solutions can be implemented at less cost than central station and transmission-dependent solutions for reliability.[47] Analysts argue that a distributed energy system, including increased use of cogeneration, is much less subject to disruption, whether from weather, terrorism, or other factors, than a centralized generation and distribution system.[48]

Since the attacks on the World Trade Centers on September 11, 2001, the vulnerability of electric systems to systematic planned attacks has come under more scrutiny. If utility transmission towers or pipelines are physically destroyed or disrupted, it can take weeks to repair them. Where an electric system is centralized and integrated, a disruption from attack at a given point can temporarily destroy large parts of the integrated network.

Several developing countries, including a few of those profiled, have been concerned that groups involved in civil unrest or terrorism could cut the transmission and distribution system or sabotage large, centralized generating stations or their fossil fuel supply to wreak economic and social deprivation. Indeed, Sri Lanka has been involved in two decades of civil unrest and succession by the *Tamil Tigers*, where the sanctity of centralized electric generation and transmission has been a continual concern, and whole regions of the country have experienced prolonged disruption. The Farabundo Marti National Liberation Front (FMNL) was able to disrupt up to 90% of El Salvador's electric production at times and produced terrorist manuals targeted just at such events.[49] The war in Bosnia-Herzegovina, where electric production and fuel supply was targeted, resulted in more than half of the generating capacity being unavailable because of direct damage, a shortage of available fuel, and half of the transmission and distribution system capacity being lost from damage and lack of maintenance.[50]

The robustness of a distributed, on-site, cogeneration-based system, likely fueled by natural gas or oil, results from

- Reliance on a larger number of small generators, no one of which is critical to supply large amounts of energy.

- Less reliance on a vulnerable centralized transmission and distribution grid.

- Reliance on the movement of natural gas fuel in the more protected underground pipeline system to the electric generation located and distributed near the demand load center, rather than reliance on more vulnerable aboveground electric transmission infrastructure to distribute electric power to the load. Gas *can* be stored in pipelines, but electricity *cannot* be stored in transmission lines, especially where they are knocked out.[51]

A large number of small units has greater collective reliability than a small number of large units, thus favoring distributed resources.[52] Distributed resources tend to fail less than centralized plants and are faster to fix.[53] A comparison of ten industrial independent power facilities and five comparably sized and constructed utility facilities indicates that the former are reliable. The mean value of availability for the five utility facilities, ranging in size from 75 to 500 MW, was 86.6%;[54] for the 10 independent power facilities the availability was 95.6%. Independent projects fired by coal average 88% to 90% availability compared with 81% availability for comparably sized, utility-owned, coal-fired plants; independent gas-fired plants show 94% to 96% availability compared to 87% to 92% for comparable gas-fired utility plants.[55]

Distributed generation can reduce losses on the transmission and distribution (T&D) system and support voltage in areas of the distribution system that suffer large drops at high load levels. With less transmission for distributed resources, there is less electron haul length and conductor and transformer heating, which increases resistance and decreases transmission system performance.[56] Thus, this lightening of transmission and distribution load allows replacing expensive voltage regulators and line upgrades. The distributed generator provides reactive-power support with both real and reactive-power injection into the system, therefore reducing current in conductors and reducing current and losses on T&D components, helping control system voltage and thereby extending the life of T&D equipment.

Distributed generation avoids the add-on costs of transmission and distribution[57] to the consumer. If the full cost of all the transformation necessary to drop voltage to household distribution levels is factored in, the true amount of system cost avoided is even higher. The typical loss of power in transmission and distribution is 7% to 10% of gross production[58] and could be as much as 20% to 40% in many developing countries.[59] Where distributed generation is effectively employed, it has been calculated that the value to the system and the customer in terms of environmental savings, reliability, engineering cost savings, electric and thermal energy value, and system deferral value can range between $300 to $1,000/kW per year or higher.[60,61] The customer does not internalize or realize all the benefits that the utility system enjoys, however; many of these benefits accrue most directly to the host utility. Seldom factored into the analysis, on average the cost to transmit electricity is more or less double the cost to ship the equivalent amount of natural gas required to make that amount of electricity.

This argues for siting generation close to load centers rather than constructing additional transmission infrastructure to move electricity long distances from centralized generators. With distributed generation fueled by fossil fuels, energy is moved more in its primary form by natural gas pipelines or oil trucks or pipelines to distributed generators, and less in its derived form as electricity, subject to transmission and distribution losses. Many of these distributed solutions in developing nations will feature renewable resources, especially in off-grid situations.

Next, we turn to the policy choices and experience with renewable and distributed energy programs in key developing Asian nations. These programs are the frontier of the global warming battle.

Notes

1. *The Boston Globe.* 2002. September 26. Business, 1.

2. Smil, V. 1999. *Energies: An illustrated guide to the biosphere and civilization*, 46–48 Cambridge: MIT Press. Phytoplankton productions are 65% to 80% of the terrestrial phytomass total, but their life span is only 1 to 5 days; Ibid. Among the most massive life forms on earth, some voluminous trees are even larger than blue whales in mass with the highest amount of phytomass; Ibid., 51; Ibid., 51.

3. Ibid., 5. This results in total solar radiation annually of 2.7×10^{24} joules. This amount of energy reaching the earth in the form of solar radiation is approximately 8,000 times more than worldwide consumption of fossil fuels and electricity; Ibid., 6.

4. Ibid., 29. The nine-hour eruption of Mount St. Helens on May 18, 1980, was approximately 1.7 EJ, or five times the total energy use of the world's annual primary commercial energy consumption.

Thermal energy rises in volcanic eruptions into the stratosphere in the form of ash clouds. From Mount St. Helens the ash rose 20 kilometers, and in the 1991 eruption of Mount Pinatubo it rose to more than 30 kilometers; Ibid., 30.

5. Villagran, E. 2000. Key drivers of improved access—Off-grid service. *Energy Services for the World's Poor: Energy and Development Report 2000*, 54.

6. United Nations. 2004. *Renewables*. (conference paper) 18.

7. Environmental and Sustainable Development Division (ESDD). 2000. *Commercialization of renewable energy technologies for sustainable development*. New York: United Nations Economic and Social Commission for Asia and the Pacific (UNESCAP). http://www.unescap.org/publications/detail.asp?id=769. (last accessed October 2004).

8. Landler, Mark. 2004. China altering energy sources. *New York Times*. June 5. B1. Ten percent would constitute 121 GW of power by 2020.

9. International Commission on Large Dams. 1998. *World Register of Dams*. Paris Database.

10. International Commission on Large Dams 2000. *Dams and Development*, 10. World Commission on Dams. During the 20th century more than $2 trillion (USD) was invested in the construction of large dams; Ibid., 12.

11. In this equation FLOW is measured in cubic meters per second, and HEAD is measured in meters.

12. Most large hydroelectric facilities are of the high head variety, with heads greater than 1,000 meters. High head plants with storage are valuable to electric utilities because they can be quickly adjusted to meet the electrical demand on a distribution system.

13. Although pumped storage sites are not net producers of electricity, their value lies in their ability to store electricity for use at a later time when peak demands are occurring. Such storage is even more valuable if integrated with intermittent sources of electricity such as solar or wind.

14. For example, there has been much concern for the ecological, cultural, and environmental impact of China's Three Gorges Dam, the world's largest planned hydroelectric power plant. Additionally, decaying vegetation submerged by flooding may give off quantities of greenhouse gases equivalent to those from other sources of electricity. If this is true, large hydroelectric facilities may be significant contributors to global warming.

15. Damming a river can alter the water amount and quality downstream, as well as preventing fish from migrating upstream to spawn. These impacts can be reduced by requiring minimum flows downstream, and by creating fish ladders, which allow fish to move upstream past the dam. Normally carried downstream to the lower reaches of a river, silt is trapped by a dam and deposited on the bed of the reservoir. This silt can slowly fill up a reservoir, decreasing the amount of water, which can be stored and used for electrical generation. The river downstream from the dam is also deprived of silt, which fertilizes the river's flood plain during high water periods.

16. When exposed to sunlight, the electrons in the crystal structure of PV cells made of crystalline silicon are knocked out. These liberated, negatively charged electrons build up a negative charge that creates an electron flow when an external path is created by a wire connecting the negative and positive areas of the cell. This flow of electrons creates an electric current in the wire, which may be used directly or stored for later use.

17. Single crystal silicon cells are manufactured by growing large crystals of silicon and cutting them into thin wafers. Single crystal silicon cells are generally the most efficient variety of PV cells, converting up to 23% of incoming solar energy into electricity. Producing single crystal silicon cells is slow and very expensive. For this reason, researchers have developed several alternatives to single crystal

silicon cells. *Thin film* PV cells made from silicon and other materials have also been developed in an attempt to lower production costs. Thin film silicon cells degrade quickly and are inefficient. Other materials, such as gallium arsenide (GaAs), copper indium diselenide ($CuInSe_2$), cadmium telluride (CdTe), and titanium dioxide (TiO_2) have all been used as thin film PV cells, with varying efficiencies and production costs.

18. Manufacturing some types of PV cells, particularly gallium arsenide cells, may involve the production of some potentially toxic substances.

19. Pace University Center for Environmental Legal Studies. 1990. *Environmental Costs of Electricity*, 36.

20. U.S. Department of Energy. CSP technologies overview, http://www.energylan.sandia.gov/sunlab/overview.html/. For descriptions of various solar-thermal systems, see U.S. Department of Energy. Technology characterization: Solar parabolic trough, http://www.energylan.sandia.gov/sunlab/PDFs/solar_trough.pdf/; U.S. Department of Energy. Technology characterization: Solar power towers, http://www.energylan.sandia.gov/sunlab/PDFs/solar_tower.pdf/. (last accessed October 2004).

21. U.S. Department of Energy. CSP technologies overview. http://www.energylan.sandia.gov/sunlab/overview.html/ (last accessed October 2004).

22. U.S. Department of Energy. Wind power today and tomorrow: An overview of the wind and hydropower technologies program, http://www.nrel.gov/docs/fy04osti/34915.pdf/. (last accessed October 2004).

23. Two main factors determine whether a crop is suitable for energy use. Good energy crops have a very high yield of dry material per unit of land, which reduces land requirements and lowers the cost of producing energy from biomass. Similarly, the amount of energy, which can be produced from a biomass crop, must be greater than the amount of energy required to grow the crop.

24. Kutscher, C. T. The status and future of geothermal electric power. National Renewable Energy Lab, http://www.nrel.gov/geothermal/pdfs/ 28204.pdf/. (last accessed October 2004).

25. Pace University, 1990 (pg. 36).

26. Energy Efficiency and Renewable Energy Network, http://www.eren.doe.gov/RE/ocean_thermal.html/.

27. Energy Efficiency and Renewable Energy Network, http://www.eren.doe.gov/RE/ocean_wave.html/.

28. Ibid.

29. Energy Efficiency and Renewable Energy Network, http://www.eren.doe.gov/RE/ocean_tidal.html.

30. Energy Efficiency and Renewable Energy Network, http://www.eren.doe.gov/consumerinfo/refbriefs/nb1.html.

31. Lovins, A and J. Heiman. 2002. Small is profitable: The hidden economic benefits of making electrical resources the right size, 303. Snowmass, CO: Rocky Mountain Institute.

32. Although microturbines are not quite as clean on emissions as new large combined-cycle, gas-fired generation units, they emit fewer pollutants than a controlled diesel engine. The emissions profile is a function both of the technology employed, the fuel used, and the efficiency of combustion. For example, A diesel engine at 40% electric efficiency is more efficient than the standard traditional steam-cycle centralized electric utility plant operating at approximately 33% efficiency.

33. Regarding global climate change, shifting land patterns are suspected by many researchers to be as significant as industrial emissions. Destruction of carbon dioxide sinks, such as forested areas, reduces the amount of carbon dioxide that can be stored, causing the remaining bare land to release less water into the atmosphere. This reduces annual rainfall, which increases local temperatures by a

significant amount. In addition, stripped land more easily releases the heat that it would otherwise store back into the atmosphere. These factors upset climate balance.

34. *World Energy Outlook,* June 1997. New York: McGraw-Hill.

35. National Acid Precipitation Assessment Program. 1989. *Interim Assessment,* vol. I.

36. These include the emission of a variety of pollutants, health impacts from these emissions, impacts on the natural environment of such emissions, and human occupational exposure or illness at the power plant work site. The primary effects on human populations are the increased risk of mortality and morbidity, including chronic illness and increased risk of chronic disease; Ibid.

37. These include waste disposal costs for residual elements of fuel and the eventual costs of decommissioning energy producing facilities; Ibid.

38. This is primarily in the form of thermal discharge from fossil fuel and nuclear power facilities, water impacts from hydroelectric dams and spillways, and leachate contamination from discharge ponds or landfills for contaminated facility water; Ibid.

39. Large hydroelectric generating stations flood upstream land; solar and wind electric production facilities create visual, aesthetic and, in some cases, television signal interference externalities; large generating facilities, particularly nuclear facilities, may adversely impact property values in the region where the facility is located; Ibid.

40. China announces plans to tackle acid rain problem, http://www.pnl.gov/china/so2.html and, for example, *China Coal News,* Shandong curbs sulfur content in burning coal. (last accessed October 2004).

41. A 15% reduction in fuel use should accompany a change from separate steam electric generator and separate low-pressure steam boiler to a steam electric cogeneration system. U.S. Government

Printing Office. 1980. *Industrial Cogeneration—What it is, how it works, its potential.* Washington, DC: U.S. Government Printing Office. EMD-80-7.

42. Greene, N., and Hammerschlag, R. 2000. Small and clean is beautiful: Exploring the emissions of distributed generation and pollution prevention policies, *Electricity Journal* 13 (June): 54, 57.

43. A diesel cogeneration system using 0.2% sulfur No. 2 oil could save approximately 0.1 lb. of SO_2 for every 100 kWh of electricity generated by the facility. U.S. Congress, Office of Technology Assessment. 1983. *Industrial and Commercial Cogeneration*, App. B.

44. Particulates are solid or liquid substances in a variety of sizes, produced primarily by stationary fuel combustion and industrial processes. Although some particulates—or particulate matter as they are commonly referred to—are noncombustible material from the original waste input, some are condensed gases from material vaporized during incineration but cooled into or onto particles. Reitze, A. W. Jr., and Davis, A. N. 1993. Regulating municipal solid waste incinerators under the Clean Air Act: History, technology and risks. *Boston College Environmental Affairs Law Review* 21: 1–21. Particulate matter is formed from noncombustible constituents in fuel or in the combustion air, products of incomplete combustion, or formation of ammonium sulfates after combustion. Typically these are unburned hydrocarbons and sulfur. Four types of emission control devices are used to control particulate ash emissions: electrostatic precipitation and filters, multitube cyclones, and wet scrubbers. Electrostatic precipitation and fabric filters remove 96% and more of particulates. Multitube cyclones are mechanical devices and are less efficient.

45. A shift in electricity generation from utility central-station conventional technologies to either gas or diesel turbine cogeneration systems will actually increase NO_x emissions, and the latter technology will also increase carbon monoxide (CO) and particulate emissions; Ibid.

46. U.S. Congress, Congressional Budget Office, *Prospects for Distributed Electricity Generation*, Executive Summary, September 1, 2003.

47. Cowart, R.,W. Shirley, R. Cowart, et al. (2002).National Renewable Energy Laboratory, *State Electricity Regulatory Policy and Distributed Resources: Distributed Resources and Electric System Reliability* 6, NREL/SR-560-32498, (October).

48. Zerriffi, H. and N. Strachan. 2002. Electricity and conflict: Advantages of a distributed system. *Electricity Journal* (Jan./Feb.): 57; Lovins and Heiman, 2002. (pg. 111).

49. Zerriffi and Strachan, 2002. (pg. 56).

50. Zerriffi and Strachan, 2002. (pg. 57 and n. 10). The generating capacity was repaired at the cost of approximately $50/kW for the first proximately 1,000 MW of repair. This is approximately 10% of the cost of new construction; Ibid.

51. Zerriffi and Strachan, 2002. (pg. 59–60).

52. Lovins and Heiman, 2002.(pg. 181). They also reduce reactive power flows by avoiding transformers. Ibid., 225.

53. Ibid., 186.

54. *See* Smith M., *Reliability, Availability, and Maintainability of Utility and Industrial Cogeneration Power Plants* 1783–1787, IEE Document No. 89CH2792.0, Oct. 1989, in Ferrey, S. 2005. *The Law of Independent Power*, 22nd ed, sec. 3:99. New York: West, 99.

55. National Independent Energy Producers (NIEP). 1992. *Negotiating Risk: Efficiency and Risk Sharing in Electric Power Markets*. Washington DC: NIEP, 9. Availability measures the percent of hours in a year that a particular project was available for operation. Capacity factor represents the ratio of kilowatt hours generated to the potential kilowatt hours (this is equal to total hours in a year multiplied by the kW capacity).

56. Lovins and Heiman, 2002.(pg. 223, 225).

57. On average, transmission and distribution costs add 25% to 50% to the delivered cost of power. U.S. Department of Energy, EIA. 2001. *Annual energy review 2000*, ch. 2, 8. Washington, DC: U.S. Government Printing Office. The average cost of transmission and distribution is deemed to be 2.4¢/kWh, adding approximately 30% to the average delivered price of electricity, which is 7.9¢/kWh; Ibid. chaps. 2, 3.

58. Ibid., table 8.1. This loss can double at certain peak summer times when power conductivity of lines diminishes. *Prospects for distributed electricity generation*, at chap. 3, 2.

59. Ramesh, M. A., *Tackling transmission and distribution losses effectively—A managerial approach*, ICECE, Oct. 29 to Nov. 1, 2003, 98, http://www.telecom.net.et/~bdu/icece2003/P029.pdf.

60. Swisher, J. 2002. Cleaner energy, greener profits: Fuel sales as cost effective distributed *Energy Resources*, 29. Snowmass, CO: Rocky Mountain Institute. Citing Lovins.

61. Frame, R., and Quinn, M. 2002. Large ITOs and traditional transmission pricing don't mix. *Electricity Journal* (Jan./Feb.): 50.

3

Overview: What Works for Renewable Power Implementation in Developing Nations

Five developing Asian nations have pioneered renewable electric energy programs to reduce the emission of greenhouse gases. Between 1993 and 2005, five nations in Asia developed small power producer (SPP) programs to promote renewable energy development in their countries. These programs have created important models and lessons.

In just a few years, these programs have made a substantial contribution of new renewable small power projects to the national energy supply: About 2% of the power supply in Sri Lanka and almost 4% in India and Thailand comes from SPP renewable energy initiatives.[1] Power development in Asia is an important global environmental and resource laboratory: Approximately 60% of all new power generation capacity financed in developing countries is in Asia.[2] Therefore, the energy resources deployed in Asia have long-term implications for global greenhouse gas emissions and environmental integrity.

These five Asian nations have different forms of government and different predominant fuel sources in their generation base (hydroelectric, coal, gas, oil). Some of the national electric systems have an integrated high-voltage transmission system, whereas others have a disintegrated or island system. However, they share some key similarities:

- All are in need of long-term increases in power generation capacity (although Thailand has a short-term current surplus).

- All have the potential for small-scale renewable energy options.

- Each country is being approached by private developers seeking to develop renewable SPP projects.

- Each system employs either deliberately or informally a standardized power purchase agreement (PPA), although it is not necessarily a neutral or consensual document in all cases.

- Although each nation recognizes the avoided cost concepts for establishing the SPP tariff, avoided cost concepts are applied differently in each nation's SPP program.

There are some common elements of tariff design. Indexation to foreign exchange rates was contemplated in most instances, but ultimately it was not implemented. Most of the programs elected not to index their tariffs to foreign exchange (Thailand's capacity payment is an exception). In some Indian states, the tariff was indexed to inflation to keep the value of international currency amid local inflation. Other countries unilaterally review and adjust the tariff at a prescribed time. This adjustment reflects changes in the energy component of the tariff, which changes when marginal fuel costs change and when foreign currency rates change, where marginal fuel is an imported commodity for the country. Although it is important, this periodic adjustment does not provide long-term predictability for project finance or equity investment.

Although these nations' programs have certain elements in common, important distinctions exist among them. The following chapters analyze in detail the programs in five nations: India, Indonesia, Sri Lanka, Thailand, and Vietnam. This analysis reflects the feedback of stakeholders in the private power process in these countries collected by the authors over a multiyear period. This evaluation highlights program differences, successes, and missteps. In addition, key provisions of the legal structure and the standardized PPAs in each country are evaluated and critiqued. The basic structure of the PPAs is analyzed:

- Elements of power sale and metering
- Allocation of risk parameters among the parties to the PPA
- Interconnection and transmission provisions
- Tariff and price design for the power sale transaction
- Parameters of SPP operation and breadth of obligation
- Dispute resolution

Each of these programs was built on prior success under the Public Utility Regulatory Policies Act (PURPA) in the United States and other Asian countries. Table 3-1 displays key comparative elements of the program design and implementation in five of the programs surveyed (Vietnam is not listed in the table because its program is in the process of being finalized; however, it is discussed later in the narrative detail about Vietnam.).

Two columns in this table should be noted. First, the middle column illustrates that two of the five profiled programs subsidize renewable energy SPPs. Thailand does so by providing a project-specific subsidy through a competitive-solicitation process. Andhra Pradesh does so by providing a tariff in excess of true avoided cost for renewable energy SPPs. The final column illustrates that some programs have an open offer to enter a PPA that the SPP may accept. Other countries that more carefully ration the PPA opportunities use a controlled solicitation of offers from prospective SPPs to award PPAs. In this situation, the SPP makes an offer or bid to the utility, which the utility may or may not accept.

Table 3-2 displays the salient comparative elements of PPA design and contractual entitlement in five of the programs surveyed.

Table 3–1: Comparative Program Overview

Country Program	Year begun	Maximum size (MW)	Premium for renewable energy	Primary fuel	Eligible PPA solicitation
Thailand	1992	<60 or <90	Yes, competitive bid	Gas	Controlled period
Indonesia	1993	<30 Java <15 other island grids	No	Renewable Energy	Controlled period
Sri Lanka	1998	<10	No	Hydro	Open offer
India: Andhra Pradesh	1995	<20 Prior <50	Yes, in tariff	Wind	Open offer
India: Tamil Nadu	1995	< 50	No	Wind	Open offer

Table 3–2: Comparative PPA Elements

Country program	Standard PPA?	Maximum years	Third-party sales	Self-service wheeling	Net meter–banking
Thailand	Yes	20–25 firm	No, under consideration	No, under consideration	Yes, if <1MW
Indonesia	Yes	20 firm 5 nonfirm	No	Yes	No
Sri Lanka	Yes	15	No	No	No
India: Andhra Pradesh	Not formally, but a de facto standardized form	20	No, previously allowed	Yes, but very expensive	Yes
India: Tamil Nadu	In development	5–15	No, previously allowed	Yes	Yes

Note the differing and evolving policies on direct SPP retail third-party sales, self-wheeling, and net metering or energy banking. Table 3–3 displays the comparative elements of the PPA tariff in the SPP programs.

Table 3–3: Comparative Tariff Elements

Country program	Avoided Cost Basis	Indexed to Foreign Currency	Periodically adjusted	Design Elements
Thailand	Yes, energy and capacity payment for firm contracts only	No	Yes	Utility purchases 65% of off-peak power
Indonesia	Yes, both energy and capacity	Yes	Yes, for changes in avoided capacity cost	Steep on-peak incentives; differentiated for each island grid
Sri Lanka	Yes, energy only; nondispatchable units receive less than full avoided energy costs	Not directly, but price linked to dollar-denominated imported oil price	Yes, and includes foreign fuel component	Calculated annually, based on three-year average imported oil price
Andhra Pradesh	Yes, not to exceed 90% of retail tariff	No	Yes	Reset every three years
Tamil Nadu	Exceeds avoided cost	No	Yes	Higher tariff for biomass than wind

The avoided cost concept and a standardized PPA are generally used. This is consistent with the PURPA requirements in the United States. Avoided cost is generally deemed the equitable point at which the utility system gets power at the opportunity cost of alternative power supply. Retail consumers are indifferent between utility supply and SPP supply if the avoided cost for this power is paid on a wholesale basis. However, not every program pays the long-term avoided cost for long-term firm power commitments. Although some programs differentiate long- and short-term avoided costs, depending on the firm or nonfirm structure of the PPA, other countries only pay short-term, energy-only avoided cost, regardless of the nature or length of the PPA obligation.

Some programs have varied or capped the avoided cost concept. The Indian state programs cap the tariff at 90% of the industrial retail tariff; the Sri Lankan program caps the tariff paid to nontransmittable providers. All programs periodically adjust the tariff. This is necessary to reflect changes in the marginal cost of fuel, a significant element of avoided energy cost. Some programs have indexed their tariffs to foreign exchange, such as the Thai program for avoided capacity; most adjust their tariffs periodically based on different criteria.

There is significant diversity in the tariff design. Indonesia provides strong incentives for the on-peak hourly delivery of SPP power, thereby decreasing the off-peak hourly prices for SPP power deliveries, so that the weighted average tariff equals avoided cost. This promotes market solutions and deemphasizes the necessity of contract remedies and default requirements. Sri Lanka employs a seasonally differentiated tariff to reflect peak system premium requirements. Tamil Nadu provides a higher tariff to base-loaded biomass projects than it does to intermittent wind projects.

Although these SPP programs are designed to equalize some of the monopsony power and create a more level playing field, the experience in several nations indicates that when the utility is the only legal buyer of SPP power, the overwhelming advantage in bargaining power that the utility enjoys must be balanced with a carefully designed and faithfully administered SPP program. Several important lessons for SPP program design and policy are revealed by analyzing these programs:

- A framework for structured SPP project development is necessary. SPPs do not spring fully born from the existing electric-sector environment. A system of law, regulation, and utility interface must exist to facilitate orderly SPP development.

- A transparent process is required to build investor, developer, and lender confidence.

- SPP programs may be initiated and sustained by an open offer to execute PPAs or by an ordered and time-limited solicitation process.
    - An open offer allows a constant, rolling development of SPPs, much like the original PURPA design in the United States.
    - An ordered solicitation can inject competitive bidding, which, if correctly administered, reduces bid prices and creates competition for the best projects and sites.
- The single-state buyer of power in most of the electric sectors can promote renewable SPPs more robustly and efficiently through (1) a program for purchasing all SPP power at its full value (avoided cost) to the wholesale system, or (2) the introduction of some combination of third-party retail sales, net metering–energy banking, or third-party wheeling.
- Utilities must interconnect with SPPs according to a straightforward procedure.
- In systems that are experiencing current and projected shortages of grid-connected power resources, payment of long-term full avoided cost (including capacity and energy) for renewable energy and small power development can accelerate the deployment of renewable facilities using "free" fuel (for example, solar, wind, flowing water, or agricultural waste) that otherwise would be wasted.
- In many systems additional subsidies, which reflect fuel diversity and environmental advantages, are used to assist higher-cost renewable energy and smaller SPP projects.
- Bidding can be strategically employed to minimize the ultimate system cost to the buyer of renewable power resource development.
- The PPA tariffs in the Indonesian, Thai, and Sri Lankan programs were designed to include capacity payments in the tariff payment for each kWh delivered, paid only if the SPP delivers power. This was designed to provide the maximum incentive to the SPP for dedicated performance and delivery of power at peak periods while not invoking any coercive penalties against the SPP for failure to perform at a set standard.

None of the five Asian programs currently allows direct third-party retail sales of power by the SPP. Thailand, which led the initial development of Asian PPAs, is considering moving toward an open market for third-party sales. This is allowed in some Indian states. This approach will likely be reevaluated in many nations as power systems are privatized, operating concessions created, and wholesale competition introduced.

Each program and its primary features, presented in chronological order of implementation, is summarized in the following sections. In the chapters that follow, each program is analyzed in detail.

Thailand Program Summary

The Thai program operates in tranches of formal solicitation by the state utility. Eligible projects mirror the requirements of the PURPA program in the United States, with size limitations up to 60 MW and, in some cases, 90 MW.[3] State subsidies are provided on a competitive-bidding basis that allows the maximum leverage of renewable SPP resources at the lowest kWh cost to the state. In this process, an amount of renewable energy subsidy is set aside by the state. Against a maximum subsidy, prospective SPPs bid for the amount of subsidy per kWh that they require to enter into a PPA with the Energy Generating Authority of Thailand (EGAT), the state utility. The SPPs are awarded subsidies beginning with the lowest SPP subsidy bid until the gross subsidy allocation is exhausted. Thus, the competitive-bidding process is employed to ration and stretch government subsidies.

This competitive-bidding process may be the most important lesson of the Thai program. The ultimate price that EGAT pays for small renewable power is a function of two price components: a fixed energy price plus a competitively bid and set renewable energy subsidy. By having potential renewable energy projects bid for the amount of subsidy they require, the least-cost (subsidy) renewable projects are selected. This rations the projects so that the most cost-effective are selected and stretches the available pool of subsidy dollars over the most megawatts of renewable power. Although the general energy price is based on utility avoided cost, the small subsidy for successful renewable energy projects provides a premium above the avoided cost. In return, the system benefits from the fuel source and supply diversity these projects provide.

As in the United States, the majority of projects are natural gas-fired cogeneration projects of independent power producers (IPPs). Both firm and nonfirm PPAs are available. The contract was designed to be indexed, but instead it is adjusted periodically to account for foreign exchange risk for capacity payments and fuel price changes for energy payments. For intermittent renewable projects, the capacity factor must be greater than 0.5 so as not to suffer a reduction in capacity payments. Thailand was the first of the Asian SPP programs, and it set a standard for successful program development.

Table 3-4 sets forth (in abbreviated format) the primary program design, tariff, and contract provisions of the Thai SPP program, including its innovative renewable subsidy incentive.

Table 3-4: Primary Elements of the Thai SPP Program

1.	**Process:** Controlled solicitation.
2.	**Maximum Size:** 60 MW (90 MW with permission).
3.	**Tariff:** Avoided cost to utility. For firm 20-year energy and capacity: Coal: $0.04 per kWh. Gas and Renewables: 2.14 baht per kWh assuming 85% capacity factor [$0.051 per kWh (2003 exchange rates).
4.	**Third-party retail sales:** No. Under consideration.
5.	**Self-wheeling:** No. Under consideration.
6.	**Energy banking:** Only for SPPs < 1 MW.
7.	**Standardized PPA:** Yes. After 2001, because of excess capacity, EGAT purchases 100% of capacity rating of kWh on peak and 65% of capacity rating kWh off-peak. Therefore, project cannot supply and be paid for rated capability during off-peak periods.
8.	**PPA term:** Firm, 5–25 years; Nonfirm, < 5 years.
9.	**Subsidy or incentives:** Competitive bidding for five-year renewable subsidy. Up to $0.009 per kWh based on lowest bids. Eight-year income tax holiday. Equipment exempt from import tax.

Indonesia Program Summary

The Indonesian program began its development in 1993. It came to involve a standardized PPA and tariff. The SPP program was designed to supply up to one-third of the national new power supply capacity additions from small renewable sources, organized into four tiers of priority for projects up to 30 MW on the primary island and half that size on smaller island grids. Because Indonesia comprises several separate and unconnected island grid systems and isolated diesel systems, this program design was nuanced and disaggregated to address avoided cost and power requirements on a regional basis.

The standardized PPA in its original design contemplated either a firm or nonfirm power sale. The incentives for firm power delivery were embodied in the tariff, with the indexation of capacity payments for foreign exchange risk, on the theory that most of the value-added cost of generating capacity would be foreign-manufactured turbines and generator sets (this program included cogeneration using fossil fuels as a lower-priority generation source). Therefore, both conventional industrial cogeneration and renewable resources were eligible for this program. This provided an innovative approach to structuring the performance obligation, whereby sanctions without a legal basis were imposed for the SPPs' failure to perform. These sanctions were accompanied by economic shortfalls for the SPP from such nonperformance. Some innovative fuel price hedging was provided for renewable power projects.

There are two lasting lessons of the Indonesian program design: (1) Disaggregated PPA provisions and tariffs can be designed to address different regional grids and requirements, and (2) PPA and tariff incentives can be designed to provide profound financial incentives for SPP delivery of power at peak times. This latter element allows the PPA to avoid typically stringent sanctions and penalties for failure to perform on peak: Market incentives are substituted for the traditional command-and-control legal sanctions.

In 1997, the Asian financial crisis ended the chances for program implementation in Indonesia just as this SPP program was being rolled out. Indonesia found itself oversubscribed in IPP power obligated on a noncompetitive basis at above-market prices amid falling demand in the face of economic crisis; consequently, it did not follow through on SPP contracts. The Indonesian State Electricity Corporation Ltd. (PLN) became a non-creditworthy party as a long-term power purchaser. Unilateral changes in the PPAs by the utility resulted in a final PPA that was not financeable without modification. Not a single megawatt of power has yet to be brought to market under this program.

Table 3–5 sets forth, in abbreviated format, the regulatory system, tariff, and contract characteristics of the Indonesian SPP system, including its tariff differentiation for peak-period power delivery.

Table 3–5: Primary Elements of the Indonesia SPP Program

1.	**Process:** Controlled solicitation. Renewables and renewable cogeneration given highest priority.
2.	**Maximum Size:** 30 MW on Java–Bali; 15 MW on 7 other island systems.

3.	**Tariff:** Avoided cost for each island system.
	For firm renewable energy and capacity:
	
	                                                        Weighted
	                     On-peak Off-peak average
	Java–Bali $0.155 $0.04 $0.059
	Other Islands $0.17 $0.05 $0.07
	$ per kWh (1995)
	
	The dramatic devaluation of the rupiah since these tarifs were calculated caused withdrawal of the tariff during the Asian financial crisis. Because of the drastic devaluation of the rupiah, the above 1995 prices are not expressed in rupiahs. 95% (of year 1) floor under renewable SPP energy price in future years (not inflation adjusted), whereas energy price can increase with marginal system fuel prices year to year; capacity price adjusted by the U.S. dollar-to-rupiah exchange rate for five years.

4.	**Third-party retail sales:** No.
5.	**Self-wheeling:** Allowed with permission.
6.	**Energy banking:** No.
7.	**Standardized PPA:** Yes.
8.	**PPA term:** Firm, 5–20 years; Nonfirm, < 5 years.
9.	**Subsidy or incentives:** Steeply incentivized on-peak tariff. Exemption from import duties and certain income taxes. Postponement of the value added tax and sales tax on luxury goods.

India Program Summary

In India, each state makes its own determinations about SPP programs, subject to federal incentives and guidance. Two representative Indian states are evaluated. Although some Indian states provide formal SPP solicitations or allow direct retail third-party sales (or both), neither of the two states evaluated here allows direct third-party sales or conducts formal project solicitations. What distinguishes the Indian SPP programs is their flexible and creative deployment of federal and state renewable energy subsidies and financing programs, which have combined to create a highly successful SPP program.

In mid-2003, the federal Electricity Act was enacted.[4] It consolidates the electricity laws and regulation embodied in the federal legislation. The act requires a license to generate and distribute electricity, except in rural areas, where no such license is required as long as the distributor follows the preestablished requirements of the central electric authority. Any generating company may construct and operate a generator without obtaining a license as long as technical grid standards are observed.[5] Transmission, distribution, and trading of electricity require a government license.[6] State commissions are directed by the act to facilitate the transmission, wheeling, and interconnection of electricity within the state.[7]

Andhra Pradesh program summary

In Andhra Pradesh state there is no formally standardized PPA, but the utility has employed a similar contract in all SPP transactions, thus making a de facto standardized PPA contract while leaving the utility extensive case-by-case discretion regarding which contracts to enter. The tariff, among the highest in India, escalates 5% annually from the base year and does not exceed 90% of the high-tension retail tariff. Moreover, the government can reset the tariff mid-contract after three years. This makes for little project certainty and is a major problem cited by project developers.

Table 3–6 sets forth, in abbreviated format, the relevant provisions of the SPP program and tariff in Andhra Pradesh, including the significant wheeling fee.

Table 3–6: Primary Elements of the Andhra Pradesh SPP Program

1.	**Process:** Open offer.
2.	**Maximum Size:** <20 MW (was <50 MW).
3.	**Tariff:** Above avoided cost to utility not to exeed 90% of industrial retail tariff. Rs. 3.32 per kWh [$0.0698 per kWh] in 2003.
4.	**Third-party retail sales:** No. (previously allowed).
5.	**Self-wheeling:** Allowed with 28% wheeling fee plus $0.01 per kWh charge.
6.	**Energy banking:** Allowed with 2% energy banking charge.
7.	**Standardized PPA:** Yes.
8.	**PPA term:** 20 years.
9.	**Subsidy or incentives:** Federal loans with 1- to 3-year repayment moratorium. 80% of capital cost can be depreciated against taxes in the first year. Grants for PV systems. Equipment exempt from sales tax.

Tamil Nadu program summary

In Tamil Nadu state no formal standardized PPA is employed, although the utility has employed the same PPA in every situation, thereby creating a de facto standardized PPA. This again gives the utility great discretion. Wheeling of power to an affiliated location—not to a third party—is permitted. An SPP is defined as any project up to 25 MW. The tariff is higher for biomass projects than for wind projects. There is no sovereign or currency risk hedge mechanism.

Table 3–7 sets forth, in abbreviated format, the salient elements of the Tamil Nadu SPP program, including its low wheeling charge.

Table 3-7: Primary Elements of the Tamil Nadu SPP Program

1.	**Process:** Open offer.
2.	**Maximum Size:** <50 MW.
3.	**Tariff:** Above avoided cost to utility not to exeed 90% of industrial retail tariff. Wind: Rs. 2.7 perr kWh [$0.057 per kWh] in 2003. Biomass: Rs. 2.88 per kWh [$0.06 per kWh] in 2003.
4.	**Third-party retail sales:** No. (previously allowed).
5.	**Self-wheeling:** Allowed with 2% wheeling charge for up to 25km transmission; 10% wheeling charge more than 25 km.
6.	**Energy banking:** Allowed with 2% banking charge.
7.	**Standardized PPA:** Yes, in final development.
8.	**PPA term:** 5–15 years.
9.	**Subsidy or incentives:** Federal loans with 1- to 3-year repayment moratorium. 80% of capital cost can be depreciated against taxes in the first year. Grants for PV systems. Equipment exempt from sales tax.

Sri Lanka Program Summary

The Sri Lanka program does not use a simultaneous solicitation for SPP bids, as in Indonesia and Thailand. Ad hoc offers are entertained by the state utility. In 2003, the program was modified to adopt a controlled-solicitation process, in which application fees and earnest money deposits are collected from PPA recipients. Letters of intent to successful bidders are valid for only six months. This prevents award recipients from attempting to prospect for hydroelectric sites that they have no resources to develop, and once they control these rights, trying to sell them to other developers.

Fifteen-year PPAs are available for projects up to 10 MW in size. All successful SPPs to date are small hydroelectric projects. The PPA is standardized, as is the tariff. Consultants provided by the World Bank assisted in the tariff development. The tariff is revised annually based on a three-year moving average fuel price, with a tariff floor of 90% of the original tariff below renewable project energy payment adjustments.

The tariff as it was eventually implemented is not indexed for foreign exchange risk, although it reflects the cost of imported fuel priced in foreign currency. It is paid in local currency on a kWh-delivered basis, which employs economic incentives rather than contract coercion to provide incentives for SPP delivery of power. With 33% inflation against the U.S. dollar during the six-year program period, even with a 75% increase in the tariff, there has been a net 40% increase in the value of the energy payments in constant foreign currency equivalent. Because of year-by-year oil price fluctuations and the resulting instability in the avoided cost, the tariff computation introduced a three-year rolling average oil price to buffer some of the swing in fuel commodity prices. The SPP projects in Sri Lanka are hydroelectric, and therefore they have a significant local component of the cost of capacity development. This tariff structure embodies the incentives for performance in the tariff itself. When the SPP delivers capacity, it is only paid for energy. Financing of projects is available with assistance from the World Bank through local banks, but stakeholders have reported a financing shortfall for SPPs that is attributable to the limited availability of long-term financing in the local financial markets.

Table 3-8 sets forth, in abbreviated format, the principal elements of the Sri Lanka SPP program, including the peak-season tariff differentiation and rolling SPP award process.

Table 3-8: Primary Elements of the Sri Lanka SPP Program

1.	**Process:** Open offer.
2.	**Maximum Size:** 10 MW.
3.	**Tariff:** Avoided cost for nondispatchable projects de facto capped not to exceed tariff paid to larger IPPs. Differentiated for wet and dry seasons. Wet season: SL Rs. 5.58 per kWh [$0.06] Dry Season: SL Rs. 6.06 per kWh [$0.062] (2003).
4.	**Third-party retail sales:** No.
5.	**Self-wheeling:** No.
6.	**Energy banking:** No.
7.	**Standardized PPA:** Yes.
8.	**PPA term:** <15 years.
9.	**Subsidy or incentives:** SPP and IPP power equipment generally exempt from import tax and enjoy tax holiday if projects are implemented under Board of Investment rules (http:/www.boi.1k).

Vietnam Program Summary

The SPP program in Vietnam is still in its initial phase of implementation. It is distinguished by its communist-socialist structure and the public rather than private companies that are the prime candidates to develop SPP projects. Nonetheless, the power development issues confronting Vietnam are similar to those of the other Asian nations: the need to develop additional power generation resources, the need to attract additional national and multinational capital to finance such expansion, and significant potential renewable electric generation resources that are available for development on a small scale in dispersed locations.

A variety of SPP and IPP proposals are in different stages of development. Oddly, in the Vietnamese system, the PPA may not be obtained until after construction of a power project is well under way. The government responsibility for all sectors of the system—development, financing, permitting, operation—substitutes for the normal checks and balances of the market. Risk is not passed on to market players; rather, all of risks ultimately rest with the government.

This system must and will adapt to the necessities of attracting international financial resources. In December 2002, Electricity of Vietnam (EVN), the national utility, committed to proceeding with SPP projects based on principles of avoided cost pricing and a standardized PPA. A major structural legislative reform of the electricity law has been in development since 1996, with enactment imminent. In 2004, Vietnam retained a new consulting team to begin program planning and design for a comprehensive SPP program. When this process is complete, Vietnam will graft onto its socialist system the sufficient market mechanisms, electric-sector reforms, and PPA and tariff provisions to drive interest in small renewable power development. It is adapting market concepts to its unique socialist government and societal structure.

In 2000, a consultant-proposed standardized PPA employed a concept of "deemed" energy as a means of paying for the capacity component of power sold separately. Deemed energy output obligates the utility to pay for the capacity component of power, regardless of whether it takes any power. This tariff concept is different from that employed by the other Asian countries. This tariff is not structured to induce maximum production and delivery of on-peak power because the payment for capacity value is not built into an on-peak energy component of the PPA tariff. This initial PPA design was never implemented. With the SPP development process restarted in 2004, all options are back on the table. The Vietnam program has the advantage of hindsight to incorporate lessons learned from the other Asian nations and international SPP experiences during the past decade.

Notes

1. Electricity Generating Authority of Thailand. 2003. www.egat.or.th/dppd/sppstatus.pdf; Energy Services Delivery Project. 2002. Grid connected mini hydro projects approved by PCIs. Colombo, Sri Lanka: Energy Services Delivery Project. http://www.energyservices.lk/mini_hydro.html. Alternatively, http://www.boi.lk lists incentive investment details for Sri Lanka. A review of programs, tender notices, and a series of reports on the India program is available at http://www.MNES.nic.in.

2. World Bank. 1998. The East Asia financial crisis—Fallout for private power projects. *Viewpoint* (August): 146. Washington, DC: World Bank.

3. Ferrey, Steven. 2000. *The new rules: A guide to electric power regulation.* Tulsa, OK: PennWell; Ferrey, Steven. 2005. *The law of independent power,* 22nd ed. St. Paul, MN: West Publishing.

4. The Electricity Act of 2003 aimed "to consolidate the laws relating to generation, transmission, distribution, trading and use of electricity...promoting competition...promotion of efficient and environmentally benign policies."

5. See the Electricity Act of 2003, sec. 7, 9. Certain conditions are imposed on the development of hydroelectric generations to ensure the highest use of water resources for competing uses. See the Electricity Act of 2003, sec. 8, 9.

6. See the Electricity Act of 2003, sec. 12, 11. Conditions may be imposed on the license. See the Electricity Act of 2003, sec. 16, 13.

7. See the Electricity Act of 2003, sec. 30, 19. An appellate tribunal also is established to handle appeals of an order of the regulatory commissions. See the Electricity Act of 2003, sec. 110, 53. State governments are authorized to constitute special courts to expedite trials of those who steal or divert electricity. See the Electricity Act of 2003, sec. 153, 68.

4 Thailand: Creative Competitive Markets in the Heart of the Tiger

Program Overview

Thailand, one of the Asian "tiger" economies, was one of the first countries in Asia to adopt a small power solicitation program. The Electricity Generation Authority of Thailand (EGAT) has installed about 22,000 MW of generation capacity, making it the largest electric system surveyed here (the Indian states are considered separate utility systems). Peak demand is 16,500 MW during April. At this level, Thailand is presently in a surplus situation, making it the only system reviewed here that is in surplus. Some of this surplus is the result of a successful IPP–SPP (independent power provider–small power provider) program. Power demand is expected to grow at about 4.6% annually.[1] EGAT is scheduled to restructure its program as a competitive retail model operating through a power pool.[2]

The Thai program was modeled on elements of the Public Utilities Regulatory Policies Act (PURPA) small power program in the United States. The roots of most of the Asian small power programs are either the U.S. PURPA program or other Asian SPP programs that were modeled on the PURPA program. Competitive bidding was introduced in the late 1980s in Maine, Massachusetts, and several other states in the United States.[3]

The monopoly state utility, EGAT, is the only entity to which an SPP may sell power in Thailand. No direct retail sales are allowed. However, such concepts are now under discussion within the regulatory authority to allow net metering or direct retail sales by SPPs, or both, in the future.

It is noteworthy that in the Thai system, competitive bidding by renewable energy SPPs is used to suppress and award subsidy payments. None of the other national systems reviewed here employ such a feature. It has been successful in minimizing the cost of subsidies and employing available subsidy funds to bring forth the maximum number of megawatts of new private power resources.

However, such a competitive system requires that there be a controlled competitive-solicitation process for SPPs. Thailand and Indonesia operate such a solicitation. By contrast, India's two states analyzed here and Sri Lanka avoid a solicitation in favor of a continually open offer to sign power purchase agreements (PPAs) and purchase power. Both systems have advantages, but it is only with competitive solicitation that competitive bidding for subsidies can be applied.

Program Design and Implementation

The following sections analyze in detail each of the key elements of the Thai program.

Solicitation of SPP participation and program mechanism

EGAT periodically announces the solicitation of additional SPP resources. This allows a competitive-bidding process to award limited subsidies to projects. In 1992 EGAT made a first request of 300 MW for SPP power. This amount was expanded to 1,444 MW in late 1995. In 1996 EGAT announced additional purchases from renewable SPPs. A formal solicitation commenced in the fall of 2001 for eventual award of contracts by the end of 2002 or early 2003; however, this process was suspended because of the power surplus situation. As of this date, no additional IPP power is being accepted, but small SPP power is still accepted.

Size and resource limitations

Eligible projects include biomass, waste, mini-hydroelectric projects, photovoltaic (PV) systems, or other renewable projects, such as wind. Conventional fossil fuels can be used for up to 25% of the fuel of such a resource annually. The eligible cogeneration technologies must use energy sequentially in a topping or a bottoming cycle. At least 10% of the annual energy output must be used in a thermal application. System efficiency of 45% must be achieved for the use of oil or natural gas in a cogeneration system. Each of these requirements mirrors the PURPA requirements in the United States in fundamental ways.

The regulations allow SPPs to deliver power for sale to EGAT up to 60 MW, although it is within EGAT's discretion to accept up to 90 MW on a case-by-case basis. The project can have a nameplate capacity greater than the limit as long as power sale is limited to the allowed capacity. Several projects at 90 MW have been accepted. However, EGAT does not contract with projects below 1 MW; these very small projects sell power output directly to one of the two national distribution companies.

Power authority role

The power authority is the sole buyer of power. The government executes the PPA and provides subsidies. The subsidies are provided competitively through a bidding process. This is an innovative feature of the Thai system. Subsidies were available in the 2001–02 solicitation process for up to five years for renewable projects of no more than 0.36 baht per kWh (US$0.01 per kWh.). The subsidies were granted under the Energy Conservation Promotion Fund Committee, which was established by the Energy Conservation Promotion Act, B.E. 2535 (1992). Two billion baht (US$50 million) was allocated to renewable project subsidies in up to 300 MW of such projects contracted after June 2000. Selected projects were required to be in commercial operation by September 2004.

The regulatory authority, the Energy Policy Office, has used competitive bidding as a tool for selecting applicants with the lowest required subsidy-adder to receive the subsidy. Additional money is allocated to other conservation and complementary programs. The application procedure for obtaining such subsidies is extremely sophisticated and documents a host of financial factors, expenses, and revenues. The average subsidy awarded is 25 baht per kWh (US$0.006 per kWh) to 31 projects totaling 513 MW.

SPPs and IPPs benefit from an eight-year tax holiday on tax applicable to net profits from project operations. There is no tax in Thailand on the sale of power output to EGAT.

Number and capacity of SPP interest and applications

EGAT has received 110 proposals.[4] Among them are 82 firm power applications for 2,800 MW, of which 50 projects for about 2,000 MW each have successfully received PPAs since 1992, when the program began. Some projects withdrew, which left about 35 firm contract projects, as well as some nonfirm projects. A project is deemed nonfirm if (1) it receives a contract of less than five years, or (2) it is an intermittent technology without capacity value. About half of the firm applications have been accepted, whereas about 70% of nonfirm applications have been accepted. Of note, most of the Thai SPP projects are not renewable projects, but are gas-fired projects. Although the program is a success and has demonstrated SPP cogeneration potential, it has not restricted participation to renewable sources. Similarly, the Indonesian program (discussed in chapter 5) promoted SPPs with a preference for (but not a restriction to) renewables compared to cogeneration.

Criteria for award

An objective and transparent scoring and evaluation process is used that reflects the factors itemized below. In this regard, it mirrors some of the most sophisticated second-generation state PURPA bidding processes in the United States.[5] The Thai program operates through a formal solicitation process. A technical submission, made in the Thai language, provides information about the following elements:

- Power plant size and capacity net of on-site use
- Renewable energy production process and fuel use
- Site location
- On-site power usage
- Feasibility studies
- Construction permits and other necessary consents
- Work plan, schedule, and project timetable
- Developer's experience
- Declaration that the information provided is true and complete

Award data

As of July 2001, the program had attracted, accepted, and obligated 44 energy projects of 1,799.9 MW, of which 156.9 MW were powered by renewables. An additional 300 MW of renewable power was being sought by EGAT.

As of the end of 2002, 71 SPPs had been accepted and obligated, with a total capacity of 2,330 MW. Thirty-five of these SPPs for 2,048 MW were firm commitments, whereas 36 projects for 282 MW were shorter-term nonfirm commitments.[6] Of these 71 projects, 23 nonfirm and 27 firm contract SPPs had entered commercial operation; the remainder were still in development. Five had not yet signed a contract. Ten of the 35 firm contracts employed exclusively renewable energy (all biomass or cofired biomass). Thirty-four of the 36 smaller nonfirm projects employed renewable resources.

From these data, it is apparent that the larger IPP projects tended to use principally conventional fossil fuels, mostly natural gas but also coal and oil, and tended to have firm PPAs. The bulk of the renewable fuel SPPs had a smaller installed and dedicated capacity and had nonfirm contracts. In this sense, the renewable energy projects were disadvantaged.

Size and type of technologies

SPPs of up to 90 MW can sell power to EGAT on a case-by-case basis, although 60 MW is the maximum size to which each SPP is entitled, according to the regulations.

The bulk of these projects are cogeneration projects, and most of these firm power projects are powered by natural gas. It is typical of these renewable SPP projects that they are cogenerating power for self-use and exporting less than the installed capacity. Many of them have an installed capacity above 90 MW and have contracted to sell EGAT 90 MW. Contract terms of 20–25 years are the norm for these larger cogeneration projects under firm contracts.

The renewable energy projects primarily combust rice husks (sometimes augmented by wood chips) and bagasse (sugar mill waste). Also represented are projects fired by wood waste and solid waste. Many of the renewable SPPs are much smaller (1–8 MW) and do not have long-term contracts, or they have nonfirm contracts that are extendable by EGAT. A few of the renewable SPPs have set 5- or 10-year contracts.

Completion ratio and reasons for failure. There has been a relatively high completion ratio.

Process transparency. The process appears to be very transparent. Scoring is done on project applications.

Stakeholder concerns. Stakeholders are concerned about the standardized PPA, which was drafted solely by EGAT. The PPA was criticized as being too simplistic and not protective of SPP interests. Many SPP and IPP projects borrowed foreign-debt capital in U.S. dollars because it carried a lower interest rate and extended longer term than available local loans. However, some sophisticated international lenders have refused to lend because the PPA for SPPs was too simple and inadequate:

- In particular, the commitment to purchase was too indefinite. The PPA has become even more critical in the current lending environment, in which loan terms have fallen to 9–10 years maximum from 12 or more years before the 1997 financial crisis.

- The financial crisis in 1997 caused a dilemma for some projects that had borrowed in foreign currency (U.S. dollars) but were receiving PPA payments in Thai baht. A fundamental restructuring required the foreign exchange indexation of capacity payments and adjustment of fuel prices.

- Concern was also raised about EGAT dispatch protocol for SPPs. Greater coordination between EGAT and the two national distribution companies was recommended. The tariff level was not criticized by developers. Third-party retail sales are allowed within industrial areas or if the government grants a concession.

- Energy banking or net metering was suggested as helpful in situations in which EGAT, now in surplus, was not accepting all power output. Energy banking could allow these projects to operate at higher levels.

Lessons

The Thai experience underscores several important lessons:

- When foreign capital is involved, indexation of capital payments to foreign currency exchange may be required.

- A controlled solicitation for SPP power can use competitive bidding to allocate subsidies for renewable power SPPs in a manner that suppresses the cost of those subsidies to the government and maximizes the amount of power brought forth by the amount of subsidies available.

- When a surplus power situation occurs, an IPP program can be suspended; however, because renewable energy SPPs use "free" fuel flows, an SPP program should be continued.

- Tax holidays are a significant stimulus to SPP development.

Power Purchase Agreements

The principal features of the Thai agreement are given in table 4-1.

Table 4-1: Features of Thai SPP PPAs

Feature	Description of SPP feature
Basic provisions	
1. Parties	Contracts are made between the SPP and the power purchaser, typically EGAT. Projects of less than 1 MW contract directly with one of two national distribution companies.
2. Milestones	The PPA contains no milestones.
3. Delivery of power	By regulation, power is delivered at the metering point.
4. Output guarantees	Where their duration is five years or less, the contracts are nonfirm without a capacity payment or a firm commitment to deliver power. For a capacity component payment, by regulation, the SPP must make a

capacity commitment of at least five years. The capacity obligation requires the SPP to supply electricity during the peak months of March–June and September–October and must supply no fewer than 7,008 hours annually of power, per the regulations, if the power source is wind, solar, or minihydro. The regulations require for waste, biomass, and tree plantations, that annual hours must be at least 4,672 annually and include March–June. By regulation, the monthly capacity factor must not be less than 0.51.

Output guarantees are in the form of limits on the time for planned maintenance and in the posting of security for contract performance. A bid bond of 100 baht per kW ($2.50 per kW) is required of applicants. The security for a bid was 500 baht per kW for some contracts. A performance bond in the amount of 5% of the total receivable capacity payment discounted to present value is required to be posted by selected applicants for the term of the contract. A Letter of Credit is a permissible means to satisfy this requirement. By regulation, all shutdowns for maintenance shall be accomplished during the off-peak months of January, February, July, August, November, and December. Maintenance shut-downs are limited by regulation to 35 days per year.

5. Engineering warranties

By regulation, the SPP must generate electricity in accord with the EGAT Regulations for the Synchronization of Generators to the System. The SPP is responsible legally for any damage to the EGAT system.

Sale elements

1. Power quantity commitment

Up to 90 MW is some instances, and up to 60 MW in size typically. EGAT recently has been accepting 100% of power output on peak, but only 65% of capacity off-peak, for a weighted average of about 80% of capacity.

2. Metering

Provisions for meter accuracy address the determination of the quantity of power sold, and procedures for redress. The meters are owned by the SPP. Accuracy is required within 2–3%. Meter accuracy in U.S. small power purchase contracts typically is required to be within a range of 0.5%–2.0% of precise accuracy.

Metering occurs at the delivery point specified by EGAT. If the SPP's meters are capable of measuring power supplied during peak, off-peak, and partial-peak periods, it receives, by regulation, time-differentiated energy payments. If not, an average energy rate is applied to all power delivered.

3. Net metering or exchange
> Not presently allowed. EGAT is taking only 65% of power output capacity during off-peak periods because it is in a temporary surplus situation. Small renewable SPPs below 1 MW contract directly with one of the two national distribution companies rather than EGAT, and for these projects net metering is permitted.
>
> No self-wheeling is permitted.

Risk allocation

1. Sovereign risk and financial assurance
> The laws of Thailand govern the interpretation of the contract. There is no sovereign guarantee. In some contracts, the SPP is required to post a bank guarantee against premature termination of the agreement.

2. Currency risk
> The financial crisis in 1997 caused a dilemma for some projects who had borrowed in a foreign currency ($), but were receiving PPA payments in Thai baht. A fundamental restructuring was required, where foreign exchange indexation of capacity payments and adjustment of fuel prices was required. The capacity payments are now adjusted for exchange rate fluctuations in the baht–U.S. dollar exchange, by a formula specified in regulation and geared to changes in the price of the fuel used.
>
> Traditionally, the energy payments are adjusted automatically for changes in the baht–U.S. dollar exchange rate, depending on the type of fuel used in the facility. This exchange was linked to changes in the price of Thai gas, Thai oil, or Japanese coal. However, as of 2001, the energy payment is adjusted, but no longer indexed. As of 2001 and thereafter, the energy payment for a firm contract was 1.49 baht per kWh ($0.034 per kWh), adjusted for changes in the price of Thai gas and not indexed to any foreign currency. In 2001 for an energy-only contract, which by definition is for a duration of five years or less, the energy payment was 1.59 baht per kWh ($0.036 per kWh), adjusted for changes in the price of Thai gas, without indexation to a foreign currency.

3. Commercial risk
> Risk is allocated implicitly to SPP. EGAT needs to take only 65% of rated capacity during off-peak periods and only 80% or more of annual SPP capacity (assuming 100% of peak-period power is taken).

4. Regulatory risk and change of law
> If there is a change of law, at the request of the aggrieved party, the parties agree to meet to attempt to resolve the issue. If no resolution is reached, the contract remains in force, and the matter is not considered to constitute a dispute for arbitration.

5. Excuse and force majeure
> Force majeure is defined to include acts of government, including seizure of the power plant, and includes otherwise fairly standard provisions of accepted international contract format.

Transmission

1. Transmission and distribution obligations
> The sale transaction occurs at the meter. Transmission on the down side of the meters is the responsibility of EGAT.

2. Interconnection arrangements
> By regulation, interconnection costs are the responsibility of the SPP prior to supplying electricity.

Tariff issues

1. Type of tariff
> EGAT incorporates an avoided cost tariff concept. Energy payments are determined from EGAT's long-term avoided energy cost for any contract with a capacity commitment, defined as five years or more. For shorter contracts, which are not eligible for capacity payments, the energy payment is determined from EGAT's short-term avoided energy cost, by regulation. These calculations approximate the fuel cost plus the O&M cost of the power plant avoided or not run during peak, off-peak, and partial-peak periods.
>
> In the 1996-vintage contracts, the energy payment was calculated as 0.87 baht per kWh ($0.035 per kWh at then exchange rates) for energy-only contracts, and somewhat less for energy in longer-term contracts, which added capacity payments of 204–285 baht per kW per month for contracts of 10–15 years (the longer the contract term, the higher the capacity payment).
>
> In 2001, the capacity payment for contracts of 10–15 years was 270 baht per kW per month. For a 20-year agreement, the capacity payment was 400 baht per kW per month. For 5–10 years, it was 217 baht per kW per month. As of 2001, the energy payment for a firm contract was 1.49 baht per kWh, adjusted for changes in the price of Thai gas, not indexed to any foreign currency.

To translate this to a representative tariff, if one assumes an 85% capacity factor for the SPP, under 2003 exchange rates, a firm PPA for a 20-year term receives a total payment (energy and capacity) of Rp.2.14 baht per kWh [$0.051 per kWh]. In 2003 for an energy-only contract, which by definition is for a duration of five years or less, the energy payment was 1.59 baht per kWh [$0.037 per kWh], adjusted for changes in the price of Thai gas, without indexation to a foreign currency. Therefore, there has been a change in the formulas for capacity and energy payments, with a simplification in the current iteration.

Traditionally, the energy payments are adjusted for changes in the baht per U.S. dollar exchange rate, depending on the type of fuel used in the facility. This exchange is linked to changes in the price of Thai gas, Thai oil, or Japanese coal. However, as of 2001, the energy payment is adjusted, but no longer automatically indexed.

2. Capacity obligations

Duration: By regulation, capacity payments are determined from EGAT's long-term avoided capacity cost for firm contracts and capacity commitments, which must be 5–25 years. The capacity payments are adjusted for exchange rate fluctuations in the baht per U.S. dollar exchange, by a formula specified in regulation, relevant to changes in the price of the fuel used. In return for capacity payments, SPPs must submit to being dispatched within the minimum 80% annual take obligation of the buyer.

Seasonal and Hourly Requirements: The SPP must supply electricity during the peak months of March–June and September–October, although no minimum seasonal capacity is specified. This eliminates certain agricultural biomass units that are not available during the peak demand seasons. In addition, to receive capacity payments, the SPP must supply no fewer than 7,008 hours of power annually (although there is no specified minimum amount of capacity during these hours) for wind, solar, fossil-fired, and minihydro, but for waste, biomass, and tree plantations, those annual hours of generation must be at least 4,672 and need only include operation during the March–June seasonal peak. The intermittent renewable resources have a more exacting hourly minimum than the base loadable renewable resources, and thus would likely not be able to qualify for capacity payments. By regulation, the monthly capacity factor must not be less than 0.51 for any project receiving a full capacity payment in a given month. Capacity payments are reduced by half if the monthly capacity factor is less than 0.51. The power factor must be between 0.85 leading and lagging, by regulation. Under the regulation, delivery of whatever amount of kW produced by a renewable resource during the seasonal peak months, and SPP availability for and delivery of some power during the minimum required hours, would satisfy these regulatory requirements. Nonetheless, most intermittent (solar, wind, run-of-river minihydro) renewable SPP projects will not qualify for capacity payments. Some base-loaded SPPs could qualify for capacity payments by operating 60% of the year including during four peak months.

3. Fuel price hedging

This is not a part of the PPA.

4. Update mechanism

The avoided fuel price of EGAT, as it changes up or down, is flowed through in periodic energy payments to the SPP, by regulation.

5. Tariff penalties for nonperformance

If the SPP is unable to supply electricity in excess of the monthly capacity factor of 0.51, the capacity payment for that month is reduced to 50% of the specified capacity payment, by regulation. Similarly, if EGAT is not able to take electricity for a period of time, those hours are subtracted from the hours used to calculate the capacity factor for purposes of capacity payments. A contract can require the SPP to supply power at stated levels in the case of EGAT need. If the SPP does not so provide such commitment when asked, for each such day 4% of the monthly capacity payment shall be withheld, by regulation. At year-end, if the SPP has not satisfied the number of hours required to supply under regulation, EGAT may recall the capacity payments already made at a rate of 0.0625% per hour for each hour of deficiency.

Performance obligations

1. Operational obligations

To the extent that the above capacity performance requirements are not satisfied by the SPP, the SPP is given 18 months to rectify the performance deficiency. If performance is not rectified, the capacity contracted for by EGAT can be unilaterally reduced to reflect actual performance, under regulation. After the midpoint of the contract term, the SPP shall have the election to reduce its contract-committed capacity with advance notice to EGAT. The SPP must be willing to commit to reduce its supply during off-peak periods (21:30–08:00 hours) to 65% of its contracted capacity upon request of EGAT. By regulation, EGAT must take 80% of annual available power. Any amount not purchased during one year is carried forward as a purchase commitment during the subsequent year, by regulation.

2. Definitions of breach
> Breach is defined in a conventional manner. The defaulting party has 15 or 90 days to remedy the default, depending on its nature. In some contracts, the SPP is required to post a bank guarantee against premature termination of the agreement.

3. Termination opportunities
> If the SPP terminates, the capacity payment is rectified with the actual term of the contract with interest. SPPs eligible for capacity payments must deposit security payments in the amount of 10% of the capacity payments expected during the first five years. This deposit is refunded at the completion or termination of the contract on terms that allow termination by the SPP.

4. Guarantees of payment and performance
> There are no outside guarantees of payment. However, late payment carries interest at 2% above the overdraft rate of the Krung Thai Bank Public Company.

5. Assignment or delegation
> Assignment is not allowed without permission of the other party, except to subsidiaries or for the purpose of financing. These are standard provisions.

6. Dispute resolution
> By regulation, arbitration is allowed to resolve disputes, with appeal to Thai courts. In the contract, arbitration is specified. Two arbitrators, one selected by each party, attempt to arbitrate disputes. They can select an umpire if they cannot agree. The arbitration proceeds under the Thai Ministry of Justice Rules in Bangkok in the Thai language. The parties may substitute by mutual agreement the Rules of the International Chamber of Commerce. A party has a right to redress in the civil courts.

Notes

1. Pichalai, Chavalit. 2002. Presentation of the Director of the Power Division, National Energy Policy Office. Yangon, Thailand: National Energy Policy Office. 27. The surplus generating capacity is expected to be absorbed by 2006 demand levels; ibid., 29.

2. Ibid., 35. There will be system operator and divestiture and privatization of assets, overseen by a new regulatory agency, National Energy Regulatory Commission (NERC), much like the Federal Energy Regulatory Commission in the United States.

3. Ferrey, Steven. 2005. *The law of independent power*, 22nd ed. St. Paul, MN: West Publishing. See chap. 9 for a discussion of the bidding schemes deployed under PURPA.

4. Pichalai, Presentation of the Director of the Power Division, 6. Thailand also purchases power from Laos and in the future from China under a long-term agreement.; ibid., 16ff.

5. For information on PURPA bidding schemes, see Ferrey, 2005. vol. I, chap. 9.

6. Electricity Generating Authority of Thailand. 2002. Internal data. Bangkok, Thailand: EGAT. The nameplate capacity of the 35 firm projects was 3,620 MW, of which 2,048 MW was committed to EGAT at no more than 90 MW per project. The nameplate capacity of the 36 nonfirm projects was 823 MW, of which 282 was committed to EGAT, with the largest being 45 MW from one of the few nonfirm, nonrenewable energy projects.

5 Indonesia: Carrots Rather than Clubs—Incentives for Peak Performance

Program Overview

In the mid-1990s Indonesia underwent significant growth in the power sector. The generating resources of the Indonesian State Electricity Corporation Ltd. (PLN), the state utility, were not sufficient to keep pace with demand growth. Approximately one-third of the installed generating capacity in Indonesia was on-site, privately owned self-generation. Because of extensive agricultural, oil and gas, mining, timber, and manufacturing interests in the nation, there was substantial possibility for renewable energy development and cogeneration. Moreover, there was a significant opportunity to entice installed "captive" generation to contract for sale to the national grid.

The government subministry, the Directory General of Electricity and Energy Development (DGEED), and PLN, with support from the World Bank, began to devise an SPP program in 1993. International consultants were hired to draft a standardized power purchase agreements (PPA) and tariffs for a small power provider (SPP) program. This was undertaken as part of a rural electrification project loan from the International Bank for Reconstruction and Development (a part of the World Bank Group). Several draft contracts and tariffs were prepared, and implementation of the final version of each became a covenant in the loan agreement between the government of Indonesia and the World Bank.

Program Design and Implementation

Solicitation of SPP participation mechanism

Based on consultant reports and a standardized tariff and PPA, the SPP program was set up to begin in 1996 (calendar year). The program was to offer a PPA in either an official English-language version or an official Indonesian-language version, at the election of the PPA and its lenders. The PPA would be standardized and not subject to negotiation. The tariff would be standardized by each region (*wilayah*) of the major island grids and isolated diesel grids. Negotiation would not be allowed on material contract or tariff issues. Choices on interconnection and commencement date would be left to the SPP. The program was announced publicly. Applications were due at a uniform time from all bidders, although, in practice, this was not uniformly honored in each region, as described later.

Renewable resources were afforded a preference in the solicitation. There were four descending tiers of project priority, with renewable energy at the top, cogeneration with renewables and with fossil fuel–fired cogeneration in the middle tiers, and conventional noncogeneration fossil fuels in the lowest tier. In operational terms, each region would award completed applications first from the top-tier renewable resources, proceeding down the hierarchy until the resources-solicitation block was filled from bids of available resources. Therefore, this "stacking" in program design accomplished a clever dynamic: It fills up the queue first with renewable resources, and then it proceeds to accept additional small power resources so that the overall SPP objectives are given full consideration.

What is important to stress is that the Indonesian program does not accept renewable projects that are less qualified than cogeneration projects. Rather, if applications are equal and both are willing to accept the avoided cost price, the program first accepts the offer of the renewable SPP. These are taken first until the allotment is filled. This makes sense, given that scarce fossil fuel resources thereby are preserved unless needed.

The agreement made by the government of Indonesia was to procure as much as one-third of future power resources under this program. Therefore, had it been implemented, renewable resources and small power resources would have had an opportunity to compete for a meaningful share of what was then a 10% or more per year power demand increase in the nation. In program design, this was ambitious. Simultaneously, though, Indonesian ministry officials were signing contracts for large, nonrenewable IPP projects on an ad hoc basis. The oversubscription of large IPPs posed an inherent conflict with the agreed SPP program, especially once the Asian financial crisis occurred in 1997 and thereafter as the SPP program was being rolled out.

Size and resource limitations

The SPP solicitation was differentiated for eight island grids in Indonesia and for isolated systems that were not connected to the transmission grid. On the islands of Java and Bali, where an integrated grid serves 75% of the country's population, projects of up to 30 MW were eligible. On other island grids, projects up of to 15 MW were eligible to submit project applications.

Power authority role

Under the officially approved program, the utility agrees to solicit and purchase 10% of the projected peak plus reserve margin on each major grid outside Java–Bali for the next eight years. On Java–Bali, the obligation is 5% of peak.

The DGEED oversaw the implementation of this project by the utility, PLN, which organized the solicitation and trained its regional offices to administer it. This decentralization resulted in significant variation in implementation in the regions.

Indonesia has a complex set of ordinances, official acts, presidential decrees, regulations, and directives that affect the provision of electricity in the country.[1] Four of these are particularly relevant to the provision of electric power by independent producers:

- Law No. 15 of 1985
- Regulation No. 10 of 1989
- Presidential Decree No. 37 of 1992
- Regulation No. 02.P/03 of 1993

If a private power enterprise sells electric power directly to the public pursuant to an Electric Power Enterprise Permit, the minister establishes a geographic service area in which it may market its power.[2] The right of industries to employ cogeneration for their own use is protected, regardless of whom their electricity supplier is.[3] Private power enterprises are required to interconnect with the utility when possible.[4] The mandate to interconnect in Indonesia is placed on the independent producer.[5]

Number and capacity of SPP interest and applications

The 1996 initial award solicitation allocated 906 MW of SPP award capacity across Java–Bali and eight regional grids on other major Indonesian islands. This represented almost 10% of the installed national generation base. Solicitations could also come from other isolated systems that were not connected to the grid. More than 70 SPPs responded by offering more than 1,100 MW in eligible projects.

Criteria for award

Applications needed to be complete and the application fee paid. Renewable resources were afforded a preference in the award criteria. There were four tiers of priority, with renewable energy at the top, renewable- and fossil fuel–fired cogeneration in the middle tiers, and conventional noncogeneration fossil fuels at the lowest tier. In other words, each region under regulation would award entitlements to sell SPP power to PLN from completed applications first from the top-tier renewable resources, proceeding down the hierarchy until the resources solicitation amount was filled with available resources. The award process fills up the queue first with renewable resources and then proceeds to accept additional small power resources in lower tiers.

Award data

From more than 70 applications representing more than 1,000 MW, only 44 projects from 27 different developers were awarded a contract, constituting 281 MW. This represents a selection rate of 26% of project capacity offered and fulfilled only 31% of PLN's 906 MW purchase obligation. Three of the eight PLN grid regions did not agree to purchase any capacity, either because they ultimately did not participate in the program, because they had summarily rejected all applications, or because they refused to sign any firm power contracts after running the process. Java–Bali, the primary island grid that serves 75% of the country's population, accepted only 13% of its required capacity allocation, constituting acceptance of only 19% of the SPP capacity it was offered. The reason stated by PLN was that most of the applications did not comply with the application protocol. An independent evaluation found little corroboration of this. More than 802 MW of project capacity offered were rejected by Java–Bali and the regional utility divisions, resulting in a significant underaward from that set forth in the World Bank loan commitments for the program.

Size and type of technologies

Of the projects that were selected and awarded contracts, which totaled 280 MW, and the 802 MW of applications rejected, the winners and losers were from the sources shown in table 5–1.

Table 5-1: Indonesia SPP Awards by Type of Energy

Source	Award winners (MW)	Award rejections (MW)	Total (MW)
Hydro	165.5	288.2	453.7
Geothermal	45.5	10	55.5
Biomass	69.5	0	69.5
Conventional fuel	0	500.7	500.7

The size of the projects accepted on Java–Bali ranged from 1.5 to 30 MW in size. In the five other PLN regions that actually made award selections (as opposed to the total that were supposed to make award selections), the size selected ranged from 1.5 to 15 MW.

The data reveal that although all biomass, most geothermal, and more than half of the hydroelectric project applications were accepted, all of the cogeneration and conventional power generation applications were rejected. This occurred even though the program was designed to accept applications up to 30 MW of all types until the program capacity was fully subscribed. There is no technical explanation as to why every nonrenewable energy project and cogeneration project was either denied, disqualified, or ignored.

Completion ratio and reasons for failure

Either PLN, the government, or both made a series of unilateral changes in the PPA that undercut project financeability. This was compounded by the 1997–98 economic recession in Indonesia and across several Southeast Asian countries and by Indonesian political turmoil.[6] The entire program was suspended in 1998 as a result of these factors, any one of which alone would have severely compromised the program rollout.

Some signs of life appeared later. In 2002, PLN agreed to purchase power from SPPs under these SPP regulations (referred to as the PSKSK Regulations in Indonesia) if the maximum capacity of the plant was no more than 1 MW. However, the purchase rate offered was 20% below PLN's average total cost of generation. Contracts are no longer offered at full avoided cost. In late 2005, international agencies began to provide support to resuscitate the SPP program, targeting the tsunami-ravaged areas of Indonesia.

Process transparency

The standardized contract was meant to offer a standardized form and to be simple and transparent. It was designed to apply to any independently financed and owned small private power enterprises anywhere in Indonesia. To achieve standardization, it had a single contract form. To make the contract more transparent and easier to understand, traditional "legalese" ("party of the first part," and so forth) was replaced with more common and straightforward terminology. To make the contract simple, effort was made to shorten it to the extent possible.

The program as designed was particularly transparent and would have succeeded, except that PLN made several last-minute changes to the PPA that made the agreements potentially not financeable. These changes also compromised the level playing field that had been carefully created in the project design and legal documents.

Stakeholder concerns

Stakeholders were concerned with both the arbitrary rejection of a significant number of applications and with the fundamental, unauthorized changes in the PPA (described in more detail later) that changed the level playing field and rendered the PPA potentially not financeable. Numerous complete and meritorious applications were rejected or ignored. Others that, in hindsight, appeared less complete were awarded contracts. Summarized succinctly by category of comment, the stakeholder concerns about this program were as follows:

Treatment of applicants

- PLN affiliates participated and were awarded contracts.
- Insiders had information.
- Some developers wanted to negotiate contract changes.
- There was resistance to implementation in the regions.
- Developers were damaged by delay and uncertainty.
- Some regions would not consider fossil fuel-fired power (25% of the awarded capacity was allowed by regulation for fossil fuel-fired power).
- No rationale for rejections was provided; some projects were rejected on technicalities or for no real reason.

Suitability and financeability of SPP contract after unauthorized revisions

- Local lenders rejected the unilaterally revised PPA.
- Lack of an official English-language version of the final PPA was problematic.
- Contract changes were seen as unfair and not financeable.
- Capacity and energy payment changes in contract tariff clauses made the contract not bankable.
- There was a need to link tariff payments to a convertible currency for more than five years.

Suitability of the tariff after unauthorized changes
- Revised tariff structure disadvantaged seasonal SPP generators.
- Revised tariff energy prices were no longer linked to oil prices posted by Pertamina (the Indonesian state oil company).
- Price stream uncertainty was a problem.

Fairness to applicants
- There was favoritism in awards.
- Large IPPs got better prices than SPPs; some developers wanted equal treatment in contract terms and tariffs.
- Most proposals on Java were rejected without cause, forcing developers to absorb the costs of the arbitrary application process.
- The program was not operated with the transparency and fairness expected.

Lessons

Despite a beginning that achieved stakeholder consensus on the PPA and tariff design, unilateral changes that the utility made in the PPA between the time of that consensus and the launch of the award process undercut the initiative. The changes made in the PPA all had the effect of making the payment terms and power sale less secure for the SPP. The elimination of an official English-language version of the final altered PPA made it difficult for others to track the changes made, required various SPP sponsors to translate the PPA back into unofficial English-language versions, and made lenders apprehensive about long-term commitments.

Even though the utility altered only a few PPA clauses, they were key clauses that established the power sale and payment scheme and allocated risk of the venture. In a shorter-form concise PPA, there is operative legal language in almost every phrase. Changing key words and phrases can fundamentally alter the legal obligations and liabilities of the parties. For several reasons, including economic and political instability in Indonesia, none of the dozens of PPA award recipients was ever able to successfully finance, construct, or operate any of the SPPs, despite some having signed contracts.

Neutral and objective PPA and tariff design is essential to successful program implementation. If one seeks to attract private capital, the program design, PPA, tariff structure, and implementation must satisfy the standards of lenders. The PPA is not infinitely fungible. On the contrary, it must satisfy a relatively precise set of criteria.

Power Purchase Agreements

The PPA as originally agreed among stakeholders

In the following analysis of the PPA, the document is analyzed as originally agreed among the stakeholders, before it was unilaterally modified by the power buyer.

The principal features of the agreements are given in table 5–2.

Table 5–2: Features of Indonesia SPP PPA before Later Modification

Feature	Description of SPP feature
Basic provisions	
1. Parties	The contract is made directly between the state utility and the SPP.
2. Milestones	The SPP has a period of two years after receiving its necessary permits to achieve commercial operation.
3. Delivery of power	The utility must accept all delivered power as long as operated pursuant to Good Utility Practices, unless the system is not able to accept power.
4. Output guarantees	The SPP pledges to commit to deliver a set amount of peak and off-peak capacity in a firm contract. Nonfirm contracts are also available. If the facility is capable of generation, it must generate and deliver power to PLN. It may not divert power to other buyers.
5. Engineering warranties	Power must be delivered at 50 Hz within 5% of nominal voltage.

Sale elements

1. Power quantity commitment

    In nonfirm contracts, there is no commitment of capacity, and energy is sold from time to time. In a firm contract for a period of years, the SPP is obligated to sell a dedicated quantity of dedicated capacity.

2. Metering

    PLN owns the metering equipment. Telemetering is required. Independent third-party calibration is required. Meters are tested annually and require accuracy within 1%. There is established a hierarchy of which set of multiple meters is employed to measure the energy and capacity sold during each billing period, cascading to secondary metering sources when the primary metering is not within accuracy parameters.

3. Net metering or exchange

    Not contemplated by the contract.

Risk allocation

1. Sovereign risk and financial assurance

    By contract, sovereign immunity is waived as a defense to suit. Otherwise, there is no limitation of sovereign risk.

2. Currency risk

    As discussed below, there is indexation to the U.S. dollar currency exchange rate for capacity payments for the first several years. This allows repayment of the capital costs borrowed in foreign currency or to purchase foreign-produced generating equipment.

3. Commercial risk

    The contract is set up so that the utility contracts for an entitlement of power, defined as a set amount of capacity plus its associated electric energy. The obligation to attempt to produce and deliver, and for the utility to take and pay for, that entitlement is absolute except for short justifiable interruptions on either side of the agreement.

4. Regulatory risk and change of law

    Although there originally was a change of legal clause covering regulatory and tax changes to allow adjustment of the price term, that clause was later removed by the utility in alterations to the PPA designed by the consultant and previously accepted by all stakeholders.

5. Excuse and force majeure

    Force majeure also is provided for both acts of God and other acts. The time limit for the maximum duration of a force majeure event is three years. This is at the most liberal extreme of the U.S. small power contracts surveyed. This provides more flexibility to attract small power producers. Force majeure is defined in a manner conventional for power sale agreements, including civil disturbance and failure of the sovereign to grant necessary permits. Failure to obtain necessary fossil fuel for the SPP or any other cause out of a party's control is also deemed to be a force majeure event. After 180 days, if not cured, the other party may elect to terminate after an additional notice of 90 days.

Transmission

1. Transmission and distribution obligations

The SPP must deliver the power at its own cost to the delivery point, and pay for all interconnection and system protective costs. Since PLN is the only entity to whom the SPP may sell power, other than its host or otherwise allowed by license, there is no obligation of the utility to transmit power.

2. Interconnection arrangements

Two options are provided for interconnection at the election of the SPP. Either the utility can build and bill the SPP for the interconnection upgrades and equipment, or the SPP can construct the interconnection equipment pursuant to utility review and standards, and then dedicate such facilities to the utility. The latter option was the one implemented by the utility. If upgrades, repairs, or modifications are later required by the utility, the SPP must implement the same at its own expense.

Tariff issues

1. Type of tariff

Energy was to be paid under all contracts at 100% of the utility's avoided cost, differentiated by region and based on actual data regarding the marginal source of generation for the utility in that region Capacity was paid at 100% of avoided capacity cost for renewable energy SPP projects, and 75% of avoided capacity cost for nonrenewable SPP projects. To share generation savings with PLN, the price to be paid to the seller is set at 75% of avoided capacity cost to PLN for other than renewable energy projects. Therefore, for firm contracts, renewable projects received a higher power purchase price.

For these projects, then, the capacity component is included in the on-peak price. This provides a substantial economic incentive to the private power producer to produce and to sell capacity at peak periods. This makes the price and contract terms simple.

2. Capacity obligations

    Capacity obligations were paid only to firm contracts, which were defined as lasting between 5 and 20 years at the option of the SPP at the time of contract formation, with an entitlement of capacity and associated electric energy pledged by the SPP to the utility. The capacity component is fixed at the time of contracting. It reflects projected avoided capacity costs for the region at the time of contract execution. The capacity component escalates annually for the first five years, and can be adjusted upward thereafter.

3. Fuel price hedging

    First, the energy component of the price changes automatically and instantaneously to reflect changes (up or down) in the price of no. 2 diesel oil, the marginal fuel for PLN. Pertamina (the Indonesian state oil company) pricing comprises this index. Second, the value of fixed and variable operation and maintenance adjusts with changes in the Indonesian consumer price index. A savings clause also is provided to dedicate the parties to finding replacement indexes if a Pertamina fuel price or a capacity price is no longer available during the term of the contract. This maintains the integrity of the contract if the chosen terms of reference are no longer available.

4. Update mechanism

    Energy prices were updated annually for both firm and nonfirm contracts. For firm contracts for renewable energy, energy prices were guaranteed not to decline below 95% of the initial first-year price, but could increase to any level if the marginal fuel cost for the utility system increased. This was done to establish a floor underneath renewable energy prices so as to facilitate their financing. Many renewable energy projects have higher capital expenditures than some conventional projects. Since the capacity and energy payments in this program are based not on the SPP's capital costs, but on the utility avoided capital (capacity) and energy costs, these renewable energy projects do not receive a higher capital cost than the utility's alternative capital cost of generation. Therefore, they must recover some of this higher capital cost, in part, through the energy payments (as well as the capacity payments) in the PPA. This floor provides assurance for a prospective lender that stability in the power sale revenue stream can be maintained to retire debt.

    The capacity payment is altered annually to hold constant the local currency–U.S. dollar exchange rate for up to three years after execution of the contract, but prior to operation, and for up to five years after operation. After this period, the capacity payment escalates only if the avoided capacity cost established yearly by the utility escalates because of annual revisions. The purpose of this is to ensure the ability to repay the cost of capital equipment purchased abroad. Such annual revisions are calculated against the initial year of subsequently executed PPAs.

5. Tariff penalties for nonperformance

The SPP is liable for direct damages to PLN for not delivering firm power entitlement during periods other than a scheduled outage. These costs include the cost of replacement power, and in a protracted situation, costs of replacement capacity.

Performance obligations

1. Operational obligations

The SPP must use its best efforts to deliver power. However, failure to deliver power for short periods, while justifying damages to the purchaser, does not rise to the level of a cause for termination. However, the tariff is structured to impose significant loss of revenue to the SPP if it does not deliver capacity on peak. Provided in this contract are the following protections of PLN:

- Seller forecasts of power to be produced and sold.
- Seller information about SPP outages.
- PLN ability not to take power when necessary.
- SPP's operation in a manner consistent with PLN standards, codes, and Good Utility Practice.
- PLN ownership of metering equipment.
- PLN rights to facility access and inspection.
- Advance notice to PLN of interruptions in sale.
- Indemnification of PLN when the independent producer owns the interconnection facilities.

2. Definitions of breach

There are no express remedies provided for breach and no explicit penalties in this contract. Although a failure to supply capacity has significant economic consequences for the seller. Moreover, no deposits or other security are required of the independent producer. There are no rights for PLN to take over the small power facility in the event that power is not provided.

Typical commercial definitions are employed. Breaches must be cured as soon as possible. A party has 45 days after notice to cure a breach, or if it requires longer, such cure must be begun within 45 days and the cure accomplished within no more than 2 years. Failure to pay within 90 days is a breach.

3. Termination opportunities
>Termination may not be made at the sole election of either party without cause. Cause for termination includes only uncured default, uncured nonpayment, or uncured force majeure.

4. Guarantees of payment and performance
>The Agreement contains no guarantees of any performance obligations.

5. Assignment or delegation
>Other than to subsidiaries for purposes of financing, the SPP may not assign or delegate its rights without the prior written consent of PLN, which may not be unreasonably withheld. A succession clause is included which has any successor to PLN assume its duties and rights regarding the contract.

6. Dispute resolution
>The purpose of the dispute resolution provision is to keep the matter out of the Indonesian court system. The parties first pledge to attempt to informally settle any dispute among themselves during a period of 60 days. If not settled, the dispute is referred to the director general of the subministry of electricity. If not then resolved within 90 days, either party may refer the dispute to the Indonesian National Board of Arbitration, which will make a final determination.

The subsequent unilateral modification of the PPA

The problem with the Indonesian program was not the program design, the tariff, or the PPA as it was originally designed and approved in the World Bank loan covenant. Later changes were made to the PPA unilaterally by PLN before it was implemented. At the end of this process, the contract was fundamentally changed and was only available in the Indonesian language. The critical changes were as follows:

- The tariff was altered to remove its set escalation provision, instead providing that it would be reviewed annually, with no indication of what would happen.

- The tariff added a capacity-factor multiplier that operated as a limitation on SPP capacity revenues so as to substantially reduce the price paid to the SPP proportionate to its achieved capacity factor (between 50 and 99% of capacity).[7]

- The limitation of a required minimum capacity factor of 50% was added.[8]
- A 120-hours-per-month limitation on SPP outage was added as a limitation on the receipt of capacity payments by the SPP.[9]
- A four-month cumulative annual threshold for the monthly limitation on SPP outage was added as a restriction on capacity payments for the SPP for the ensuing year (in addition to forfeiting monthly capacity payments).[10]
- The tariff was changed to allow alteration of the energy price each year or to reduce the energy price to 95% of the full avoided cost in the year of contract execution; previously, full avoided cost pursuant to a set formula was provided.[11]
- PLN was able to refuse taking firm power under several contingencies at its sole election.[12]
- Alterations against the SPP were made in the change of law and force majeure provisions.[13]
- Sovereign risk and enforceability were compromised for the SPP.[14]

Some of these elements are now present in other programs, such as a variable capacity factor that is adjusted based on performance (Thailand) and an annual buyer adjustment of the tariff without a contractually established formula. Not one of these changes was necessarily fatal alone, but altogether they tilted the balance of power among the parties within the PPA and made it very difficult to obtain conventional financing. With a stable economic and political climate, the PPA could have been reoriented as necessary and the program continued.

It is worth emphasizing, however, that the Asian financial crisis in 1997 and thereafter undermined this SPP program. The changes in government also undercut the necessary program continuity. With an oversubscription of large IPP power by the government, executed at ad hoc rates and with ad hoc PPAs, there was surplus power under contract when the financial crisis and recession hit.[15] The SPP program had just completed the award of contracts but project construction had not begun, it became a marginal casualty, and never moved forward.

Notes

1. Sequentially the laws, acts, regulations, and directives affecting the provision of electricity are as follows:

    - Ordinance of 1890 (O.G. no. 19, 1890)
    - Ordinance of 1934 (O.G. no. 63, 1934)
    - Act No. 19 of 1960 (O.G. no. 5, 1989, 1960)
    - Government Regulation no. 19 of 1965 (O.G. No. 39)
    - Act No. 9 of 1969 (O.G. No. 40, 1969, 2904, 1969)
    - Government Regulation no. 11 of 1969 (O.G. No. 20, 1969)
    - Government Regulation no. 30 of 1970 (O.G. No. 42, 1970)
    - Government Regulation no. 18 of 1972 (superseded by Regulation No. 17 of 1990)
    - Presidential Decree no. 59 of 1978
    - Government Regulation no. 36 of 1979
    - Government Regulation no. 54 of 1981 (superseded by Regulation no. 17 of 1990)
    - Decision of Director General for Energy and Electricity (DGENE) no. 0236/47/500/1983
    - Decision of DGENE no. 0237/47/500/1983
    - Law no. 15 of 1985
    - Government Regulation no. 10 of 1989
    - Government Regulation no. 17 of 1990
    - Presidential Decree no. 21 of 1990
    - Presidential Decree no. 37 of 1992
    - Government Regulation no. 02.P/03 of 1993

2. Indonesia. Ministry of Mining and Energy. 1993. Regulation no. 02.P/03/M.PE/1993. Jakarta: Ministry of Mining and Energy. This regulation was promulgated to execute Presidential Decree no. 37/1992.

3. Ibid., art. 45. "Cogeneration" is defined as a process by which all thermal energy produced by or recovered from a turbine is used for an industrial production process. This diverges from the definition in U.S. federal law, which (1) does not require that all thermal energy be used (only a percentage of total energy must be usable thermal energy), and (2) does not require that use must be in an industrial production process (space conditioning or other useful employment of thermal energy is allowed).

4. Ibid., art. 46. The equipment specifications of the Indonesian Electricity Standards and any other standards approved by the minister govern the interconnection.

5. U.S. Code of Federal Regulations. Regulations under Sections 201 and 210 of the Public Utility Regulatory Policies Act of 1978 with Regard to Small Power Production and Cogeneration; Electric Utility Obligations; Obligation to Interconnect. *Code of Federal Regulations* 18:292.303(c)(1). Washington, DC: Government Printing Office.

6. World Bank. 1998. The East Asia financial crisis—Fallout for private power projects. *Viewpoint,* note no. 146. Washington, DC: World Bank. Indonesia was hit hardest by the 1997 financial crisis among Southeast Asian developing nations.

7. Revised Power Purchase Agreement (PPA), art. 2, cl. 3, Capacity Factor Limitation. A new clause was added that allows the Indonesian State Electricity Corporation Ltd (PLN) to set or restrict the capacity factor that governs the facility's payment eligibility annually.

8. A capacity factor is a new concept that is added to work a reduction in the amount of compensation received by the small power provider (SPP) to reflect its actual capacity factor of delivery.

9. Revised PPA, art. 5, cl. 3, Loss of Monthly Capacity Payments. This new provision states that if there are 120 hours of nondelivery outside the scheduled maintenance and repair (which now requires prearrangement with PLN but was not previously the case), the SPP loses all capacity payments for that month. If the SPP loses 16% of its operating time in a month (equivalent to five days) either

continuously or intermittently, as a result of machine failure, failure to get fuel, or even failure of the PLN transmission system, 100% of the capacity payments are lost for that month. Achieving an 83% monthly availability factor, which by itself is not seen as problematic for most utility systems, results in loss of capacity payments. A capacity factor is already factored into the tariff. This provides PLN the ability to avoid routine delivery by the transmission and distribution system for which it is responsible.

10. Revised PPA, art. 5, cl. 4, Loss of Annual Capacity Payments. Another new provision provides that if the 120-hour failure occurs four or more times in any contract year, the SPP not only loses all capacity payments in those four (or more) months, but also loses the capacity payments for the entire ensuing year.

11. Revised PPA, art. 5, Deletion of Continuation Clause. A provision in the original contract—which provides that if an index fails of its essential purpose, it will be replaced by the parties so that the contract can continue to function as intended—has been deleted entirely.

12. Revised PPA, art. 1, Definitions: Firm and Non-Firm Capacity. The definition of firm capacity was altered so that PLN can refuse the delivery of energy from an SPP if it has no need for the energy after the contract is formed and in force. The definition of PLN's entitlement was changed from the capacity and energy *committed* to the capacity and energy *sold*. This change shifts significant power to the power purchaser to not take power.

Revised PPA, art. 1, cl. 6, Power Purchase. In the original version, PLN could temporarily interrupt accepting power only when it was necessary and consistent with good industry standards. In the revised version, that stoppage can occur for any reason that PLN deems necessary.

In both the original and utility-revised versions of the PPA, art. 2, cl. 7, Purchase Cessation. The original version contained a provision [Art. 2 (g)] designed to protect the power seller by limiting PLN's refusal to take power to situations that are absolutely necessary and with maximum possible notice. This has been replaced by two provisions that (1) allow PLN to refuse to take power (as long as it does not count against the capacity factor concept that PLN has added to this contract), and (2) state that if the seller does not cut power delivery when asked (presumably for any reason) by PLN, then PLN can cause that amount of power delivery to be cut for the power seller.

Revised PPA, art. 5, cl. 1, Tender of Power. The obligation of the SPP power seller has been changed from *tendering* to *delivering* power to PLN. Delivery is the actual provision and surrender of the power. Tendering is evidencing a present capacity and capability to deliver power. A seller of power may wish to tender power and receive a corresponding tender of payment from the power buyer to be assured that payment will be forthcoming. If delivery of power is required, then to prevent breaching the contract, the seller must deliver power before receiving an assurance that payment for the power will be forthcoming.

13. Revised PPA, art. 5, cl. 18, Change of Law. A provision was added indicating that SPPs are susceptible to any change of law or tax, without adjustment in the SPP contract.

    Revised PPA, art. 6, cl. 1, Force Majeure. A provision omits the failure of the PLN system as a force majeure event that the SPP could claim in failing to deliver power. The following changes also were made in the revised PPA to the force majeure provisions:

    - Acts of God, fire, explosion, failure of the fuel supplier, and uncontrollable events are omitted as events of force majeure.
    - By changing the language, the failure of PLN to pay money owed to the SPP now could be an event of force majeure, whereas the previous contract did not recognize this as an event of force majeure.
    - There is an added a requirement that a relevant public authority provide an explanation of the force majeure.

14. See the revised PPA, art. 3, cl. 2, Enforceability: Sovereign Risk. A provision was deleted that provided that contesting the basic enforceability of the SPP contract constituted an event of default.

The requirement to give written notice of default to the defaulting party and to allow its lenders to cure the default within a reasonable time was deleted. The definition of the commencement date of operation was changed to give PLN the power to agree as to when this occurred for the SPP project, with no requirement that PLN agree at any particular time or circumstance. In the former version, this occurred automatically when the facility began operation.

See also the revised PPA, art. 5, cl. 12, Dispute Resolution. Whereas the previous agreement stated that amounts due could be contested for up to one year, that provision was reduced to seven days. There is no provision for regular calibration of the meters, given other changes that have been made to the standardized contract.

15. World Bank. 1998. The East Asia financial crisis—Fallout for private power projects. *Viewpoint*, 5. Washington, DC: World Bank. Wholesale tariffs for large, private IPP power projects ranged from 5 and 8.5 cents per kWh during the long term. At the same time, the retail rates that PLN charged customers were just slightly more than 7 cents per kWh. Indonesia was hit hardest by the 1997 financial crisis among Southeast Asian developing nations. PLN responded by trying to lock payment rates at the old rupiah–dollar conversion rate, which was only one-third the rate after the rupiah fell during the financial crisis.

6. India: State Power in a Federalist System

Program Overview

In India, about 42%–44% of the rural population has access to electricity.[1] There is significant variation in access to electricity across various groups in the different states of India.[2] Rural electrification on an ambitious schedule will require that private-sector, renewable, and off-grid resources be encouraged and deployed.

In mid-2003, the federal Electricity Act of 2003 was enacted.[3] It consolidates the electricity laws and regulation embodied in the federal legislation. The act requires a license to generate and distribute electricity, except in rural areas, where no such license is required as long as the distributor follows the preestablished requirements of the central electric authority. Any generating company may construct and operate a generator without obtaining a license as long as technical grid standards are observed.[4] Transmission, distribution, and trading of electricity require a government license.[5] State commissions are directed by the act to facilitate the transmission, wheeling, and interconnection of electricity within the state.[6]

India has become a major player in renewable generation and private-sector power development. India is the 10th-largest developer of small hydroelectric facilities and the 5th-largest developer of wind power, as well as the 5th-largest producer of photovoltaic (PV) systems in the world.[7] In India, state electricity boards provide electric power. Much of the authority for electricity policy resides at the state rather than federal level. A number of states have small power producer (SPP) programs.

There are two particularly novel elements of the Indian system: First, it operates through the close coordination and integration of programs and policies at both the state government and federal levels. Second, it has the most complex system of subsidies and financing arrangements for renewable energy investments of any of the Asian nations.

Subsidies

In 2000–01, the federal government's Ministry of Power initiated a program called the Accelerated Power Development Program, which provides financial assistance to the states for modernization programs. To receive assistance, the beneficiary state must undertake certain reforms that promote a rational power administration at the state level and promote renewable energy technologies.

At the federal level, the Ministry of Non-Conventional Energy Sources (MNES) provides grants and subsidizes renewable power so as to create a level playing field for various energy sources. The MNES can subsidize state-owned (not private) renewable energy projects through grants to state governments, electricity boards, and transmission companies:

- Up to 60% of the cost of wind turbine equipment

- Up to Rs 15,000 per kW (US$320 per kW) for microhydroelectric power and double this amount for minihydroelectric power up to 3 MW (not to exceed 50% of project cost)

- Up to 100% of site survey expenses.

For private or public projects, the MNES can provide grants to subsidize interest rates:

- Grant subsidies of Indian Renewable Energy Development Agency (IREDA) interest rates by a reduction of 1%–3% for biomass and bagasse (sugar mill waste) projects

- Reduction of IREDA interest rates for waste-fueled projects down to 7.5%

- Reduction of IREDA interest rates for PV projects down to 5%.[8]

Certain renewable energy technologies also receive preferential federal tax treatment. Wind energy components pay a much lesser customs import duty than assembled wind generators, whereas PV systems are exempt from excise duty. The depreciation for solar devices, biogas equipment, wind turbines, and agricultural and municipal waste conversion equipment is allowed accelerated depreciation.[9] The MNES has issued guidelines for tariffs to be offered by state electricity boards to purchase renewable energy SPP power. This federal recommendation is not based on an avoided cost or long-term marginal cost calculation; rather, these recommendations provide a common starting point, but states vary the actual tariffs by as much as 50%.

The MNES estimates that, to date, the potential and realization of renewable energy sources in India is as set forth in Table 6–1.[10] The government is considering setting a national goal to achieve a minimum 10% renewable energy share by 2012.[11] IREDA, an MNES-owned renewable energy financing agency, has received funding assistance from the World Bank, the government of the Netherlands, the Asian Development Bank, the Danish International Development Agency, the Bank for Reconstruction in Germany, and the Overseas Economic Cooperation Fund in Japan.[12] Table 6–1 scales the realization of various renewable energy development in India, as of 2003, against its potential.

Table 6–1: Realization of Renewable Projects in India, as of 2003

Technology	Potential (MW)	Realization (MW)
Wind	45,000	1,267
Small hydro up to 25 MW	15,000	1,341
Biomass power	19,500	308
Biomass cogeneration	3,500	273
Urban and industrial waste	1,700	15
Photovoltaics	Significant	47

Several of the states provide other state-level renewable energy subsidies, such as a cogeneration subsidy, a sales tax exemption on generation equipment, or no tax or duty for a number of years on electricity sales in areas where direct third-party sales are allowed.[13] Renewable resource development has proceeded most vigorously not where the renewable energy regime was best, but where the state policies were most favorable. Today, a majority of renewable energy development in India is undertaken by private and nongovernmental organizations.

Financing

The significant penetration of renewable SPPs in certain Indian states is partly a function of the programmatic encouragement of such projects at the state level, but it is also promoted by a series of subsidies and financing originating at the federal level (though often administered by the states). The MNES provides direct subsidies, and a PV subsidy has been advanced by the Global Environment Facility to IREDA, which has enjoyed the infusion of more than US$350 million in international support.[14] The MNES develops promotional policies for renewable energy wheeling,[15] energy banking,[16] and third-party retail sales,[17] which are recommended to the states for implementation.

Arising from a base of almost zero financing, during the past decade IREDA and many other lenders have created a liquid market for renewable and SPP power developments in India. IREDA is an independent, specialized public-sector lending agency under the MNES incorporated in 1987. IREDA loans to private SPPs and IPPs must be secured by a loan guarantee from a commercial bank, by a secured equitable mortgage hypothecated against all immovable project assets plus a subordinated lien (second to other project lenders) against all existing and future movable project assets plus project developer guarantees, or by irrevocable guarantees by Indian public financial institutions. The mortgage applies even to projects on government-owned land, or, if not possible, obtains a letter of assurance from the state government owning the land. Until such mortgage is executed, the borrower is charged extra points on the loan interest rates. A bank guarantee or pledge of fixed deposit receipts must be provided for at least 10% of the IREDA loan amount. Less demanding loan provisions as to interest rate and loan origination fees are imposed on SPP project entities owned by women, ex-servicemen, and handicapped individuals to promote their participation in the sector, as well as on SPP projects located in certain economic development zones or remote areas. IREDA will provide SPP generation equipment financing and equipment export promotion. These loans are eligible for prepayment by the borrower. About 10% of IREDA loans are co-lent with a commercial bank. Most loan applications to IREDA are brought forward by a particular state nodal agency working with IREDA.[18] For SPP and independent power producer (IPP) renewable energy projects, IREDA loans offer a variety of terms. The terms of IREDA loans are set forth in table 6–2.

Table 6–2: Terms of IREDA Renewable Project Loans, 2003

Technology	Interest rate (%)	Loan term (years)	Moratorium on repayment (years)	Maximum amount of loan
Hydro				
<1 MW	13	10	3	75% of project cost or 100% equipment cost
1–5 MW	13.25	10	3	70% of project cost or 100% equipment cost
5–15 MW	13.5	10	3	Same
15–25 MW	13.75	10	3	Same
Wind				
Leased	12.75	10	1	70% of project cost
Ownership	12.5	10	1	70% of project cost or 100% of equipment cost
BOOT basis	11.75	10	1	70% of project cost
Off-shore demo	11.5	10	2	70% of project cost
Biomass cogeneration	13.25	10	3	70% of project cost
Waste to energy				
Industrial waste	13.5	8	2	70% of project cost
Municipal waste < 3 MW	12	10	3	
3–6 MW	12.5	10	3	70% of project cost
Fuel cells	11.5	8	2	70% of project cost
Photovoltaic	14*	10	1–2	80–85% of project cost

* With MNES subsidies, the effective interest rate for PV is 5%.
IREDA 2002b, pp. 54–63.

The most favorable terms (interest rate, repayment moratoria, percentage of project debt extended) are provided to the most expensive per-installed-unit projects (such as PV, offshore wind demonstration, fuel cells, and smaller installations). More cost-effective technologies are given IREDA loans that offer interest rates close to commercial lending rates. IREDA provides an essential function in extending debt to a sector of the economy that has had difficulty accessing local credit on sufficient terms. In this way, not only has IREDA lending facilitated essential renewable energy project development by funneling multilateral and bilateral funding, it also has provided a lending model for local commercial banks in India. Therefore, it has served as a catalyst, as well as a primary lender to the sector. For example, small hydroelectric power development is the most significant renewable energy resource currently deployed in the world—and the densest renewable resource. Four hundred million tons of agricultural waste is produced annually in India; biomass electric generation is CO_2 neutral. Also, 27.4 million tons of municipal waste per year and 12,145 million liters per day of liquid municipal waste (sewage) are generated in India. India has more than 400 sugar mills capable of generating an additional estimated 3,500 MW of surplus power from biomass sugar cane waste, if developed. IREDA has sanctioned the following number of projects by technology (table 6-3):[19]

Table 6–3: Projects sanctioned by IREDA

Technology	Projects sanctioned	MW sanctioned	Projects installed	MW installed
Small hydro	101	311	48	147
Wind		651		1625
Biomass	28	166		88
Biomass cogeneration	31	445		227

More than 3,400 MW of renewable projects were in operation by the end of 2001, from a base of about 100 MW in 1992 before these initiatives began. This increased the percentage of renewable energy in India from 0.13% in 1992 to 3.4% in 2001. Wind accounts for almost half of this renewable capacity; 26 interconnected PV projects are in service. In addition, there are 400 sugar mills in which renewable energy cogeneration is possible, as well as a potential of 16,000 additional biomass opportunities. It is estimated there is 15,000 MW of small hydroelectric potential (of 25 MW each or larger).[20] Thirty-five small hydroelectric projects of about 118 MW, 27 wind projects of 87 MW, and 78 PV projects of 2 MW have been financed by IREDA. Renewable wind projects enjoy 100% tax depreciation at the national level and subsidies in some states such as the deferral of sales tax payments. These incentives are being scaled back as the renewable energy sector matures.[21] IREDA has sanctioned loans of about Rs 5,000 crore (US$1.05 billion) and disbursed about half of this to support 844 MW of renewable power development.[22] About 25% of the renewable energy development in India has been assisted by IREDA financing.[23] Currently, wind developers report they are able to borrow commercially at a lower cost than IREDA rates.[24]

As part of this analysis, we will evaluate the program in two Indian states: Andhra Pradesh and Tamil Nadu. Andhra Pradesh is the most advanced in installing wind capacity, whereas Tamil Nadu is the fourth most successful in wind capacity.[25] The experience in each of these states illustrates the creative application of financing sources and state subsidies to successfully promote renewable SPPs, which have become a hallmark of the Indian program. Although each of these states has an advanced SPP program, neither permits direct third-party sales. Some states in India have cut back on the ability for third-party wheeling or net metering because the utilities were losing their best-to-serve customers. Each of these two states will be examined separately.

Andhra Pradesh

The following sections detail the relevant provisions of the SPP program and tariff in Andhra Pradesh.

Program design and implementation

Andhra Pradesh has a system of more than 7,000 MW that is short of capacity to serve the existing demand. Its demand curve is relatively flat, with a slight peak from 6 p.m. to 9 p.m. daily. During this afternoon peak, system voltage drops. Inflation is running at about 6% annually, and the rupee depreciates against the dollar by about a little less than that amount annually.

Solicitation of SPP participation mechanism

In some other Indian states, such as Himachal Pradesh, Haryana, and Uttar Pradesh, formal bid packages for SPPs were prepared and sold to potential project developers. In Andhra Pradesh, there was no formal solicitation or bidding; project developers could submit requests that were independently judged. Applications for contracts can be submitted at any time on a standardized form. The transmission utility, the Transmission Corporation of Andhra Pradesh Limited (APTransco), which is distinct from the Andhra Pradesh State Electricity Board (APSEB, the distributor), unilaterally makes the initial decision as to whether to enter a power purchase agreement (PPA) with an SPP developer. The same entity that makes this decision selects the rate it is willing to pay to purchase the power from such a facility and negotiates a PPA with the project.

Size and resource limitations

SPP project eligibility was originally established for projects up to 50 MW, but this has since been lowered to a maximum of 20 MW. Any renewable or unconventional waste technology project is eligible.

The amount of power sought is distinguished by size. The amount of participation by wind and small hydroelectric projects is unlimited in the program design. There is, however, limited wind-resource potential, and suitable hydroelectric sites are limited by environmental considerations in Andhra Pradesh. Cogeneration is considered a high-priority resource. For the purposes of economic and social policy, rice husk biomass projects in each district of the state are limited to employing no more than 25% of the rice husk residue. Rice husks are also employed for cooking fuel in households and for the insulation of ice slabs for transporting fish. Diverting a larger percentage of risk husks to small power generation would cause the price of husks for residential and other commercial purposes to increase.

Power authority role

The state of Andhra Pradesh is promoting renewable and unconventional power generation resources pursuant to the Andhra Pradesh Electricity Reform Act of 1998 (Act No. 30 of 1998). The act empowered the Andhra Pradesh Electricity Regulatory Commission (APERC) and created the Non-Conventional Energy Development Corporation of Andhra Pradesh (NEDCAP), which operates under the Principal Secretary for Energy of the government of Andhra Pradesh. It makes grants for renewable energy projects with funds provided by the MNES through NEDCAP, the local nodal agency.

To facilitate SPP financing, NEDCAP coordinates the projects of IREDA, which is a federal government agency. The World Bank and the Global Environment Facility support IREDA lending. Since 1997, IREDA has supported 221 projects for Rs 11.16 billion (Rs 1,116 crore or US$231.5 million) in Andhra Pradesh.[26] IREDA operates with the dual goals of renewable technology promotion and financing. Funding at conventional interest rates is available from IREDA, the financing wing of the MNES. Prior clearance from NEDCAP is required for IREDA financing.

APERC issued GOMs No. 93 (1997) and GOMs No. 112 (1998), as extended, which provide the rates paid to unconventional energy projects. The 1997 order was designed to encourage renewable energy resources by establishing a power purchase price of Rs 2.25 per kWh, escalating at 5% per year. It also allows third-party sales and energy banking. However, the 1998 order clarified that, in the absence of the grant of a license pursuant to the Electricity Duty Act of 1939, SPPs cannot make third-party sales.

Number and capacity of SPP interest and applications

There are no reliable data on the number of applicants. APTransco is willing to commit up to 2,222 MW of total unconventional energy.

Criteria for award

The applications are evaluated by APTransco. If they are deemed viable, APTransco negotiates an individual power purchase rate and contract. Appeals by the SPP applicants have gone to the APERC and to the courts. Some cases are pending before the courts as to the rates set.

Award data

Andhra Pradesh has approved the construction of 1,013 MW of unconventional generation. This is scaled against the potential reported in Table 6–4.

Size and type of technologies

See table 6-4.

Table 6-4: Andhra Pradesh Renewable Project Status, 2003

Technology SPP	Capacity (MW)			
	Projects approved	Projects complete	Projects at finance close	Potential capacity
Wind	283	92	10	745
Biomass	345	81.5	110.7	627
Bagasse cogeneration	210	49.5	75.5	250
Municipal waste	23.6	0	0	40
Industrial waste	36	1.5	4	135
Small hydroelectric	95	69	30.4	1,252

Completion ratio and reasons for failure

As of 2004, 189 MW of capacity was in operation; the remainder was in various stages of development. Because of the lack of a standardized contract or tariff, some projects have ended up under appeal at the APERC or in the courts. Projects must secure land clearance from the municipal body and any required environmental clearance from the State Pollution Control Board.

Process transparency

There is no standardized contract. General guidelines for the contract are issued by APERC and APTransco, but contracts are negotiated case by case. There is a generalized interconnection agreement, which is not required to be standardized. The PPAs are not publicly available. This adds to the inability to achieve consistency or to minimize legal and other transactions costs.

Stakeholder concerns

SPP developers have been disappointed with the results of the process in some cases. Because there is no formal standardized contract, individual negotiation occurs with the state utility monopoly to determine the contract terms. Similarly, the state utility determines the purchase rate it will offer the SPP. The utility has attempted to abrogate existing PPA tariffs. Aggrieved SPPs have appealed to APERC, which has set the rate on appeal. The decisions of APERC have been appealed by aggrieved SPPs to the Supreme Court. Wind-project developers report that they require a constant, unchangeable, standardized PPA to allow them to finance new projects during a 5- to 10-year loan period. The 80% depreciation deduction for the first year, together with point coordination function incentives, is enough to make wind projects viable with a set PPA that is enforced by the regulator.

Lessons

The role of the independent regulator is essential. APTransco is attempting to void some prior PPAs and tariffs at PPA rates that it now claims are above its average cost of generation. In discussions, it has indicated that it would prefer not to purchase power from SPPs.[27] APTransco has staked out a position opposed to certain expansion of renewable energy SPPs.[28] These issues are now before APERC and in court litigation. In one case, the court disallowed the SPP the opportunity to make a third-party retail sale at Rs 5.2 per kWh, but it upheld the obligation of APTransco to pay the stipulated Rs 3.32 SPP power purchase price if it objected to a third-party retail sale by the SPP.

APERC is considering SPP tariff revisions that would affect all existing and future PPAs. The criteria for setting the future tariff would be the differentiation of SPP profitability by technology and by necessary rates of return on investment. This would diverge from avoided cost principles. The resulting uncertainty has caused project development to halt. APERC has required APTransco to bring its current accounts payable into a more prompt payment for SPP power—within 15 days.

Wind development in Andhra Pradesh has gone through cycles. Originally encouraged by tax incentives (100% first-year depreciation), wind SPPs were often located at mediocre wind sites. In 1997, the marginal corporate income tax rate was reduced; an alternative minimum tax was introduced, and some of these project became less viable. With delays in payment for power purchases by APTransco, many of these projects have had difficulty, some going into default on their loans.

The top priority for developers is the restoration of third-party retail sale authority for SPPs. APTransco has strongly resisted this because of the reality of "cream skimming" its best customers and because it claims first access to customers—that is, the retail rate it is allowed to charge is subsidized and does not recover all system costs.

Power purchase agreements

Because there are no standardized contracts in Andhra Pradesh, unlike the four other countries in this comparison, there is no standardized PPA to evaluate. Therefore, a representative PPA must be evaluated. Each Indian state may operate its own SPP initiative if it so chooses. For the two Indian states evaluated here, PPAs for bagasse cogeneration SPPs of approximately 20 MW are discussed.

This PPA is the least specific and developed of those in the five countries evaluated here. This contract is very friendly to the SPP in that it there is no performance obligation, and energy can be delivered at the will of the SPP. The principal features of the agreements are given in Table 6–5.

Table 6–5: Features of Andhra Pradesh SPP PPA

Feature	Description of SPP feature
Basic provisions	
1. Parties	The contract is made between the SPP and the utility, APTransco. The agreement is for 20 years.
2. Milestones	Under its financial guarantee agreement with NEDCAP, the SPP is required to achieve financial closure within six months of signing a memorandum of understanding with NEDCAP and is required to begin construction within 15 months, of signing a memorandum of understanding with NEDCAP.
3. Delivery of power	Delivery occurs at the interconnection point.
4. Output guarantees	Operation of the project is totally within the control of the SPP. Power must be accepted by APTransco except for system emergency reasons. There is no warranty to deliver any energy or capacity by the SPP.
5. Engineering warranties	Power must be delivered at 50 cycles per second (–5% or +3%).

Sale elements

1. Power quantity commitment

    No commitment whatsoever, either for energy or capacity, or both, is made by the SPP for power sale.

2. Metering

    A pair of bidirectional meters are installed by the utility. Check meters are then installed by the SPP. Meter accuracy is checked twice yearly; meters are calibrated yearly. Where the primary meters do not register accurately, the check meters are utilized for billing purposes. Detailed meter testing is specified.

3. Net metering or exchange

    Power can be banked for up to 12 months at a cost of 2% of the energy banked. This is fairly typical of other states in India. Wind produces power during summer months, which is peak period, making energy banking not a primary issue in this state.

    In 2002, rates were Rs. 3.32 per kWh (2.25 per kWh in 1993–94 rupees, at 5% escalation annually) given in U.S. dollars. This rate is next revised in 2003. This rate is fairly typical of other states, but among the highest. The SPP is entitled to standby and backup power supply at the High Tension tariff.

    In the past, direct third-party retail sales were allowed, but in 1999 they were suspended indefinitely and the prior arrangements annulled by APERC at the request of APTransco. The APERC issued an order, which prohibited direct third-party sales. Power may be sold only to APTransco at the rates that they prescribe. In other states, such as Karnataka, Madhya Pradesh, Maharashtra, and Rajastan, third-party SPP sales are permitted. Uttar Pradesh, with commission approval, allows third-party wheeling without charge for the wheeler, for power sold at the same rate as that for centralized power supply. Other states, such as West Bengal, Tamil Nadu; Gujarat, and Kerala, do not allow third-party sales of private power. Tariffs are under development for third-party, high tension wheeling of power from IPPs to third-party consumers of high-voltage power.

    APTransco has allowed some wheeling in-kind. APTransco historically charged a 2% fee for wheeling. Other states typically charge 2–5%. Now, APTransco charges a 28.4% in-kind charge for the wheeling, which is the systemwide power line loss factor, plus Rs. 0.50 per kWh ($0.009 per kWh) paid in cash. At these high rates, SPPs are discouraged from wheeling power and are economically compelled to sell power to the utility.

Risk allocation

1. Sovereign risk and financial assurance

    No provision to protect against this risk is provided.

2. Currency risk
    No provision to protect against this risk is provided.
3. Commercial risk
    No provision to protect against this risk is provided.
4. Regulatory risk and change of law
    No provision to protect against this risk is provided. The SPP remains responsible for any later imposed taxes or levies. Any modification of the agreement can only be made if approved by APERC.
5. Excuse and force majeure
    No provision for either force majeure or excuse for failure to deliver is made. Since there is no obligation to deliver, there is no delivery obligation on the SPP.

Transmission

1. Transmission and distribution obligations
    The utility will transmit power to a remote location for the generator. Initially, a 2% wheeling charge was charged. The wheeling charge currently is 28.4%, to reflect what APTransco assesses as systemwide transmission grid losses, irrespective of the distance traveled. This is implemented by requiring the generator to put in 128.4% of the generation they transmit. In addition, the generator is charged a wheeling fee of Rs. 0.5 per kWh. As a practical matter, this has financially eliminated any generator transmission.
2. Interconnection arrangements
    Interconnection is designed, installed, owned, and operated by APTransco, the costs for which are reimbursed by the SPP. However, in the case of wind developers, the developer pays Rs. 1 million per MW ($21,044 per MW). A charge of Rs. 0.10 per kWh (Rs. 0.10 per kWh) ($0.002 per kWh) is charged to handle reactive power for wind generators.

Tariff issues

1. Type of tariff
    The tariff for this nonfirm power is subject to unilateral change at any time by the state government and APERC. It was set at an energy rate of Rs. 2.25 per kWh ($0.04 per kWh), escalating at 5% per annum from the base year. As of 2003, this has increased the tariff to Rs. 3.32 per kWh ($0.06 per kWh), with future revisions in March 2003 and thereafter. APERC is considering a future SPP tariff differentiated by type of SPP technology that would apply to all existing and future SPPs. This would allow the ability to differentially promote select technologies.

This is one of the highest tariffs of any of the states. APTransco claims its marginal cost of power is Rs. 2.3 per kWh ($0.042 per kWh), which understates the true cost without state subsidy. The average commercial retail tariff is in the range of Rs. 4–5.5 per kWh ($0.08–0.11 per kWh). Rates for agricultural consumers are heavily cross-subsidized.

2. Capacity obligations

There is no capacity obligation and no payment for capacity. All renewable projects are nonfirm. Only for nonrenewable larger IPP PPAs, are capacity payments included in some instances. As this is a PPA for a cogeneration facility whose production varies seasonally, the amount of power delivered changes. Excess power without a capacity commitment is sold to the utility.

3. Fuel price hedging

There is no fuel price component.

4. Update mechanism

The originally set tariff is set for only the first 3 years of the 20-year term. Thereafter, APERC may review the power sale price.

5. Tariff penalties for nonperformance

There are no tariff incentives or penalties for nonperformance. The tariff mechanism is not utilized to effectuate the incentives of the agreement.

Performance obligations

1. Operational obligations

The SPP must operate the project subject to prudent utility practices.

2. Definitions of breach

Breach is not defined in the PPA.

3. Termination opportunities

There are no provisions for termination prior to the term of the PPA.

4. Guarantees of payment and performance

If payment is made on or before the due date, the SPP receives a 1% discount and rebate credit on the next bill. If late, 14% per year interest is added.

5. Assignment or delegation

Assignment of either party must have the prior consent of the other party, which cannot be unreasonably withheld.

6. Dispute resolution

APTransco must discuss with the SPP any disputes on bills.

Tamil Nadu

The Tamil Nadu system is more than 7,000 MW and needs surplus capacity. It can serve a daytime peak of about 6,800 MW on a typical day. About 86% of villages, but less than half of the households, are electrified. Line losses have been estimated by the Tamil Nadu Electricity Board (TNEB) to be about 16.5%. During evening off-peak times, the TNEB has asked biomass SPPs to back down to 90% of capacity. The state of Tamil Nadu has more than half of all of India's wind turbine capacity and a significant percentage of its biomass projects.

Solicitation of SPP participation mechanism

Neither the government nor the TNEB solicit any proposals. SPPs can offer proposals and make applications. Decisions are made on a case-by-case basis.

Size and resource limitations

An SPP size limit of 50 MW is imposed. Multiple generator sets at the same location can be packaged into separate applications to effectively exceed this limit at the site. If the power is wheeled for one's own consumption at a remote location rather than sold to the grid, more than 50 MW is allowed.

Power authority role

The TNEB will respond to SPP proposals, but it does solicit proposals. The state has discontinued the state tax exemptions that were previously applied to certain renewable energy equipment. All environmental and siting approvals are the responsibility of the SPP.

Number and capacity of SPP interest and applications

There is no limitation on the amount of SPPs that can be developed at their own initiative. It is assumed there would be some practical limitation should there ever be a brisk development of renewables, merely to keep the system balanced between intermittent and base-loaded resources.

Criteria for award

There are no quantitative criteria for evaluation. If the TNEB appraises the SPP project as viable, a PPA is executed.

Award data

As of September 2002, the awards reported in Table 6–6 were in place.

Table 6–6: Tamil Nadu SPP PPAs awarded

Energy Source	Output (MW)
Wind	894
Bagasse	186
Other biomass	13
PV	2
Small hydro	13

Size and type of technologies

Most of the SPP projects are wind, bagasse, cogeneration, biomass gasifier, and PV. Tamil Nadu is a major locus of wind power generation. There is a wind turbine testing laboratory in the state.

Completion ratio and reasons for failure

The TNEB claims to accept all viable projects.

Process transparency

There is supposed to be an electricity regulator to deal with disputes. This position was not filled, however, until 2002.

Stakeholder concerns

Developers are concerned that third-party sales are not allowed. The risk of a single buyer is significant. The need for third-party retail sales is the paramount concern of SPP developers.

The avoided cost concept has never been implemented: Utilities often look at the cost of buying power from other state-subsidized utilities as their cost, even though this is artificially depressed. These shadow subsidies make it impossible to see the true avoided cost in the market without a tariff study.

An electricity regulator was established only in mid-2002 and does not yet have an established track record for the handling of complaints or disputes. A model PPA was drafted and circulated in 2002 for possible adoption. The fact that it is a model does not mean that it is standardized for adoption. Wind generators were concerned that they would required to be certified by European labs as to their capacity. There is now a testing laboratory in Tamil Nadu, and self-certification is allowed.

Lessons

The PPAs reviewed in both Tamil Nadu and Andhra Pradesh are the least complex of all PPAs in the five nations reviewed. This may be because the PPAs were only for the sale of surplus energy at the sole discretion of the SPP. Therefore, these PPAs outline only the most rudimentary concepts. The rights of the parties are much less secure in these situations. Where one party is a monopoly with state sanction and the only entity to whom power may be sold, this lack of specificity works to the disadvantage of the SPP. The PPAs in the other countries evaluated were not lengthy documents, although the addition of a few more pages makes the position of the parties much more secure when a complete contact is used for SPP power sale. In Tamil Nadu, about 80% of wind-generated power is used captively by the owners, and about 20% is sold to TNEB.

Power purchase agreements

Because there are no standardized contracts in Tamil Nadu, unlike the four other countries in this comparison, there is no standardized PPA to evaluate. A model PPA has been circulated for comment, but it has not yet been implemented or used. The model PPA is not dissimilar, although it is distinct from, the actual existing and executed PPAs analyzed here. For the two Indian states evaluated here, PPAs for bagasse cogeneration SPPs of approximately 20 MW are discussed.

The SPPs are required to make reasonable efforts to operate during peak hours, although no penalty exists for this failure. The model tariff would escalate 5% per year for the first 10 years of the contract and would be subject to mutual negotiation thereafter. However, constraints are imposed on tariff escalation. A major problem is that the PPAs allow the utility unilaterally to alter its terms or the SPP tariff at any time during the term of the contract. This is a major impediment, and SPP developers report they execute these PPAs under protest.

This Tamil Nadu and Andhra Pradesh PPAs are the least specific and developed of those in the five countries evaluated here. These nonfirm Indian contracts are very friendly to the SPP in that there is no performance obligation and energy can be delivered at the will of the SPP. However, the price for this is that the SPP is paid only for energy and gets no capacity credit, even if reliable capacity is provided by the SPP. The contract summarized is a contract for excess energy produced by a cogeneration facility whose output varies seasonally. The principal features of the agreements are given in table 6–7.

Table 6–7: Features of Tamil Nadu SPP PPAs

Feature	Description of SPP feature
Basic provisions	
1. Parties	The contract is made between the SPP and the utility, the Tamil Nadu Electricity Board ("Board"). The agreement is only for surplus power that the SPP may elect to deliver. The particular agreement reviewed ranged between 5 and 15 years. It is subject to periodic renewal or renegotiation.
2. Milestones	None.
3. Delivery of power	Delivery occurs at the interconnection point. At the end of each month, the SPP must forecast to the Board its likely deliveries during the upcoming month.
4. Output guarantees	Operation of the project is totally within the control of the SPP. Power must be accepted by the Board, except for force majeure reasons. There is no warranty to deliver any energy or capacity by the SPP. The term for

biomass projects is 15 years, whereas for wind power SPPs there is no term, although the utility reports that it informally will honor these PPAs for 20 years.

5. Engineering warranties

The SPP designs and installs at its own expense its own protective equipment for parallel operation.

Sale elements

1. Power quantity commitment

No commitment whatsoever, either for energy or capacity, or both, is made by the SPP for power sale. The utility will only purchase surplus power. In the hours between 11 p.m. and 6 a.m., the utility now requires the SPP (except for bagasse cogeneration which require the steam production) to back down some of the power sold to the grid.

2. Metering

A pair of bidirectional meters is installed, but in this case by the SPP. Check meters are then installed by the SPP. Meter accuracy is checked twice yearly; meters are calibrated yearly. Where the primary meters do not register accurately, the check meters are utilized for billing purposes. Detailed meter testing is specified.

3. Net metering or exchange

Energy banking is allowed for a 5% in-kind energy charge. The SPP is entitled to standby and backup power supply at the high-tension tariff.

The SPP is allowed to wheel power over the Board's grid to its affiliated entities. When a wind project is developed, for example, a special project company can be created to be owned in shares by several companies, each of which wheels power from the wind turbine, sited to maximize wind capture, to its factory or load center. Within 25 km of it generation source, 2% is deducted for line losses; at more than 25 km, the wheeling charge is 10%. Other than this arrangement, there are no direct third-party sales currently, although it was briefly allowed in the past. Power may be sold only to the Board at the rates that they prescribe. An earlier provision to allow third-party sales was discontinued.

Risk allocation

1. Sovereign risk and financial assurance

No provision to protect against this risk is provided.

2. Currency risk

No provision to protect against this risk is provided. In the model PPA that is circulating, the Board would provide a letter of credit from a commercial bank in favor of the SPP to serve as a surety for one month's expected power payments from the Board.

3. Commercial risk
> No provision to protect against this risk is provided. Developers are required to have control of the site and a purchase order for equipment prior to signing the PPA with the utility.

4. Regulatory risk and change of law
> No provision to protect against this risk is provided. A major problem is that the PPAs allow the utility to alter its terms or the SPP tariff at any time during the term of the contract. This is a major impediment, and SPP developers report that they execute these PPAs under protest.

5. Excuse and force majeure
> A relatively weak force majeure provision that includes rebellion, riot, and natural disaster, is included. In the model PPA that has circulated, the force majeure provision is somewhat stronger, including work stoppages, fire, and loss of license.

Transmission

1. Transmission and distribution obligations
> Power can be wheeled to affiliates, as discussed above.

2. Interconnection arrangements
> There is no standardized interconnection agreement. Interconnection arrangements vary. Interconnection at lower voltages is designed, installed, owned, and operated by the SPP at its own cost. For higher-voltage interconnections, the SPP is required to deposit the cost of the work with the Board, which performs the work. For bagasse-fueled cogeneration, the Board bears the costs itself and performs the work. For a wind project, an interconnection charge of Rs. 15.75 lakhs per MW of installed capacity is paid by the SPP for the interconnection.[1]

Tariff issues

1. Type of tariff
> The tariff for wind was set at an energy rate of Rs. 2.7 per kWh ($0.057 per kWh), after September 2001, with no annual escalation. It is designed to remain unchanged for five years until 2006. For biomass (including bagasse), the tariff is Rs. 2.88 per kWh ($0.06 per kWh), representing an escalated base-year 2001 rate, limited not to exceed 90% of the high-tension transmission tariff (that is, Rs. 2.88 per kWh). This will escalate at 90% of the escalation of the high-tension transmission tariff. The retail tariff is state subsidized and does not recover sufficient revenue to cover costs. The retail tariff is expected to increase at a significant rate in future years. The rationale for this tariff difference is that wind is intermittent and does not supply any capacity value reliably. In addition, the wind regime in Tamil Nadu is better than the wind regime in Andhra Pradesh, and wind machines seem able to operate at the lower tariff.

2. Capacity obligations
>There is no capacity obligation and no payment for capacity. Because this is a PPA for a cogeneration facility whose production varies seasonally, the amount of power delivered changes. Excess power without a capacity commitment is sold to the utility.

3. Fuel price hedging
>There is no fuel price component.

4. Update mechanism
>Note the PPA itself allows the utility to change the tariff or any other clause at will. In 2002, for the first time a regulator was appointed.
>The tariff will be periodically reviewed by the regulator in Tamil Nadu.

5. Tariff penalties for nonperformance
>There are no tariff incentives or penalties for nonperformance. The tariff mechanism is not utilized to effectuate the incentives of the agreement.

Performance obligations

1. Operational obligations
>The utility reserves the right not to take power when not needed, and in the PPA the SPP agrees to back down generation during off-peak periods. Any excess above the amount of energy requested by the Board is not paid for by the Board. It does this now by requiring SPPs to back down sold power output at daily off-peak evening times.

2. Definitions of breach
>Breach is not defined in the PPA.

3. Termination opportunities
>Termination is allowed by the Board if any technical condition of the Board is not followed.

4. Guarantees of payment and performance
>None.

5. Assignment and delegation
>No contractual limitations.

6. Dispute resolution
>Disputes as to power quantity or payment arise, they are referred to the government Chief Electrical Inspectorate to resolve. In the model PPA circulated, arbitration is mandatory under the Arbitration and Conciliation Act of 1996. The arbitrator's decision is final and enforceable by the courts under the laws of India. In the model PPA, all consequential and special damages are waived.

---

[1] One lakh is 100,000, such that 15.75 lakhs is Rs. 1,575,000 ($33,500) per megawatt [$33 per kW].

Notes

1. India. Ministry of Power. 2003. *Discussion paper on rural electrification policies*. New Delhi: Government of India. This leaves almost 80 million rural households without access to electricity. Notwithstanding, as of March 2003, 87% of inhabited Indian villages were declared electrified.

2. Ibid., 1, 8. There is a government policy of universal access to electricity by 2012.

3. India. The Electricity Act of 2003 aimed "to consolidate the laws relating to generation, transmission, distribution, trading and use of electricity…promoting competition…promotion of efficient and environmentally benign policies."

4. Ibid., sec. 7, 9. Certain conditions are imposed on the development of hydroelectric generations to ensure the highest use of water resources for competing uses; see the ibid., sec. 8, 9.

5. Ibid., sec. 12, 11. Conditions may be imposed on the license; see the Electricity Act of 2003, sec. 16, 13.

6. Ibid., sec. 30, 19. An appellate tribunal is also established to handle appeals of an order of the regulatory commissions; see ibid., sec. 110, 53. State governments are authorized to constitute special courts to expedite the trials of those who steal or divert electricity; see ibid., sec. 153, 68.

7. World Bank. 2002. *Confidential data and reports*. Washington, DC: World Bank. India now exports its wind and PV technology.

8. India. Ministry of Non-Conventional Energy Sources (MNES). 2001. *Renewable energy in India: Business opportunities*. New Delhi: Government of India.

9. Ibid.

10. Ibid., 1.

11. MNES. 2002. *Annual report 2001–2002*. New Delhi: Government of India. It also has a goal to provide electric power to remote villages through the deployment of stand-alone renewable energy systems.

12. MNES. *Renewable energy in India*. 42. Assistance is provided to projects in the form of debt instruments up to 80% of the project cost or 90% of the equipment costs. Interest rates vary at levels below 15% for a period of up to 10 years with a three-year repayment moratorium.

13. MNES. *Renewable energy in India: Business opportunities*. 59, 63.

14. World Bank. 2002. *Confidential data and reports*. Washington, DC: World Bank.; Government of India, Indian Renewable Energy Development Agency (IREDA). 2001. *Annual report 2000–2001*. New Delhi: Government of India; IREDA. 2002. *Renewable energy, energy efficiency financing guidelines*. New Delhi: Government of India. Funds to IREDA have been provided by the World Bank, GEF, IBRD, ADB, KfW, and the Japanese Bank for International Cooperation. The types of financing offered to private developers by IREDA include project or equipment financing, and umbrella financing Where international or bilateral funds are loaned by IREDA, the donor requirements and stipulated procedures are imposed on the borrower as conditions of the loan.

15. MNES. 2001. *Renewable energy in India: Business opportunities*. 59. Of nine Indian states surveyed, all have implemented the MNES recommendation for energy wheeling at a charge of 2% of wind energy wheeled. Two states raised the charge to 12.5% and 20% (and one state did not allow wheeling) for biomass power, with a 2% charge to compensate for line loses.

16. MNES. 2001. *Renewable energy in India: Business opportunities*. 59. Of nine Indian states surveyed, eight allowed energy banking for wind power for a period of 6–12 months. For biomass power, energy banking was allowed for 8–12 months in six states and for 24 months in Uttar Pradesh state. Energy banking carried a 2% charge assessed as a deduction in net power wheeling in Andhra Pradesh and Tamil Nadu states.

17. Ibid. Of nine Indian states surveyed, five allowed direct third-party retail sales of power for wind, whereas seven allowed direct sales for biomass energy (Andhra Pradesh has since suspended this privilege).

18. IREDA. 2002. *Renewable energy, energy efficiency financing guidelines.* 7, 18–19, 20, 24–28, 38.

19. IREDA. 2002. *Biomass power generation guidelines for loan assistance.* New Delhi: Government of India; IREDA. 2002. Waste to energy guidelines for loan assistance. New Delhi: Government of India.

20. MNES. *Annual report 2000–2001*. 7; World Bank. 2000. Confidential data and reports. Washington, DC: World Bank.

21. World Bank. 2000. Confidential data and reports. Washington, DC: World Bank.

22. MNES, *Annual report 2000–2001*, 9.

23. IREDA, *Renewable energy, energy efficiency financing guidelines*, 2.

24. Personal interview with wind power developers. Hydabad, India. December 12, 2002.

25. MNES, *Renewable energy in India*, 8.

26. IREDA, *Annual report 2000–2001*, 32.

27. Personal communication in Hyderabad, Andhra Pradesh, December 11, 2002. it's the marginal cost of new combined cycle gas-fired power is claimed to be about Rs 1.8 per kWh (US$0.035 per kWh). The utility also is concerned that it cannot control where SPPs choose to site projects.

28. Ibid.

7

Sri Lanka: 21st-Century Conversion of Ancient Renewable Technologies

Program Overview

Sri Lanka is a single island nation. Unlike Indonesia, there are no multiple grids. However, unlike mainland nations, power cannot be imported or exported. Sri Lanka is a hydroelectric-based power system. By the end of the 19th century, almost 50 small hydro plants were operating in the country.[1] Thus, Sri Lanka has a history of dispersed renewable energy facilities, many for tea plantation shaft power for processing.

The state utility, the Ceylon Electricity Board (CEB), maintains a monopoly on retail power sales. Independent power providers (IPPs) are allowed to own and operate power projects that sell power to the CEB or consume the power for self-use. Large IPPs are able to negotiate individual PPAs with the CEB.

In 1999, the national utility grid in Sri Lanka had 1,600 MW of installed generation, supplying 5,800 GWh annually. The bulk of this installed generation was large hydroelectric projects. In the mid-1990s, consultants recruited by the government with funding from the World Bank determined a tariff structure for small power provider (SPP) development and a neutral power purchase agreement (PPA). This formed the backbone of SPP procurement.

A key factor in the development of SPPs in Sri Lanka is the well-integrated grid system throughout much of the island. This is the result of the distribution network reaching the plantations in the mountainous interior regions of the country. This allows the development of SPPs, even in relatively remote locations, and the transmission of power to the centralized grid and other retail users.

Program Design and Implementation

Solicitation of SPP participation mechanism

Although Sri Lanka encourages SPP development, the country has no formal competitive solicitation process for SPP development. The government signed some letters of intent (LOIs) with several developers who were interested in developing some of the hydroelectric sites. Because of impediments to development, these early developers dropped out of the process. No formal solicitation process has been put forth by the government or the CEB for SPPs (other than a notice in the newspapers advertising the CEB's interest in purchasing power).

The CEB signs LOIs with SPP developers on an ad hoc basis. This is true even if the developer does not have site control. Traditionally, the CEB made no detailed technical or financial evaluation before entering an LOI. Therefore, LOIs did not substantiate viable and feasible projects. LOIs traditionally could be obtained by speculators to block the competitive development of particular sites. About 70 MW of LOIs have been executed by the CEB with small power developers.

Starting in 2003, the CEB amended its procedure to require a non-refundable initial processing and application fee of Rs 1,000 (US$10), along with a prefeasibility study itemizing the hydroelectric water resource to be used and demonstrating the financial capability of the SPP sponsor. The CEB evaluates this application, whether an LOI has already issued for the site, CEB plans, and the transmission and distribution system's capability to absorb power input to determine whether to issue an LOI. The LOI is only valid for a period of six months. A performance bond of Rs 2 million per MW (US$20,661 per MW) of planned plant capacity and a processing fee of Rs 100,000 (US$1,033) is required from the LOI recipient. Unlike prior practice, the LOI is not transferable.

All of these reforms to the transferability, earnest money deposit, and limited duration are designed to prevent speculators without the resources to develop and finance an SPP from tying up renewable energy sites. These are important reforms to ensure a liquid and competitive market, especially when the resource is site specific and involves government resources, such as hydroelectric projects on government-controlled waterways. Some developers have opined that the six-month window is insufficient to obtain all of the necessary government permits and to sign the required PPA to prevent the LOI from lapsing. These developers suggested that it would be more appropriate for them to submit a performance bond at the time they receive the LOI, without a six-month maximum time to sign the PPA, with the bond refundable if they do not receive all of the necessary government permits to enable them to execute the PPA.

Size and resource limitations

Small hydroelectric and other renewable energy developers of facilities no larger than 10 MW are allowed to sign a standardized PPA with the CEB. The term of the PPA varies up to 15 years.

Power authority role

According to stakeholders and evaluators, the process changed during the mid-1990s, when the World Bank funded a consulting team to devise a standardized small power PPA and an associated tariff based on avoided cost principles that would be applicable to all projects up to 10 MW.[2] The standardized PPA eliminated market uncertainty and, along with a methodology to determine the avoided cost of the CEB, facilitated the development of the small power sector.[3] In addition, the World Bank provided funds from the International Development Agency to the government to encourage commercial and development banks to lend on longer terms at conventional rates to SPP developers. Asian Development Bank funds also were made available to develop the plantation sector of the economy and could benefit plantation SPPs. These multilateral bank funds have found their way to Sri Lanka commercial banks that are lending to SPP projects going forward. SPPs have been project financed with these funds.

When developers invest more than US$750,000, they are allowed to import equipment duty free and pay a concessionary corporate income tax rate of 15% annually for seven years. When total investment exceeds US$7.5 million, complete tax exemption is given for 10 years. The Central Environmental Authority must approve each project, the regional government authority must approve the use of water from the stream to generate power, and the chief electrical inspector must license the generation and sale of power.

Number and capacity of SPP interest and applications

Data were not available on the number of LOI applications. A substantial small hydroelectric capacity exists in Sri Lanka. One authority estimates this as 250–300 MW at sites where facilities of up to 10 MW could be developed, plus an additional 200–250 MW at sites where 10–25 MW facilities could be developed.[4] This adds up to about 500 MW of financially viable small power hydroelectric sites. However, the CEB estimates this to be a much smaller potential, and nothing approaching this potential is under active development.

Criteria for award

Contracts have been awarded by the CEB without a formal solicitation process based on viability.

Award data

Initially, with the assistance of a nongovernmental organization, two small existing electric facilities serving plantations were connected to the grid. This involved moderate additional expense because the facility already existed and demonstrated an initial CEB–SPP relationship. An 800 kW greenfield project followed suit. All of these projects used hydroelectric technology. The first project under the new initiative connected in 1997. In total, by the end of 2002, 13 SPPs with 43.3 MW had interconnected to the CEB. This represented about 2.5% of electric power production capacity in the grid.

Size and type of technologies

All of the awarded projects were small hydroelectric project, with the single exception of one small cogeneration facility.

Completion ratio and reasons for failure

A completion ratio cannot be calculated because precise data on applications are not available. As of 2002, about 43 MW of the SPPs were grid connected and generating power. An additional 20 MW were under construction, involving many of the same developers. Financing was provided with World Bank funds through the Energy Services Delivery Project and later by the Renewable Energy for Rural Economic Development Project. Approximately 200 MW of potential projects at 55 sites received LOIs, and applications for another 60 MW at about 40 sites were filed. The number of projects actually completed was a small fraction of those that had sought approval to develop.

There is no formal government policy to promote SPP development. The price paid for power is not based on any capacity payment. Only an energy component is paid, and this value fluctuates annually. Thus, there is no long-term certainty for the tariff.

Process transparency

The lack of a formal program has discouraged some early interest from international developers, who also were frustrated in navigating all of the intricacies of the siting and approval process. At least one stakeholder reported that the ongoing, unilateral changes in the tariff undercut the confidence of some of the SPP developers and eliminated technologies other than small hydroelectric SPPs from consideration. Nevertheless, domestic developers appear to have mastered the process. The 2003 requirement of a monetary deposit for limited duration of and nontransferability of LOIs will create a process to screen speculation from the SPP program.

Stakeholder concerns

Before the introduction of the standard PPA and tariff and the provision of loan funds by the World Bank, there were major impediments to financing SPPs. The local banks worked with relatively short-term deposits and did not have experience with credit facilities to project finance SPPs. Therefore, debt capital was not available in sufficient quantity or for sufficient terms to finance SPP development.

There was international interest in Sri Lankan hydropower prior to the World Bank program in that country, when the government made several large hydroelectric sites available. Several foreign developers have dropped out of project development from frustration. This has benefited domestic developers, who also have some cost advantages on small projects. There has been a dispute between developers and the CEB regarding the annual process for tariff update. Stakeholders have complained in the past that the tariff does not provide a capacity value and does not fully provide for avoided energy cost. There now seems to be less acrimony about the annual tariff update process, although it is still not set out in an objective formula.

In July 2005, CEB increased the tariff paid for small power producers to U.S. $0.06/kwh. This was important in two regards. First, it changed the tariff going forward for all existing and new SPPs to a U.S. dollar-denominated tariff from one previously denominated in local currency. This tariff will now hold its value in dollar terms, and will make it easier to finance projects in dollar-denominated loans from international lenders or finance acquisition of foreign project equipment and parts. Second, it represented an increase of 10% in the tariff level.

Lessons

Even where profitable renewable energy SPP sites exist, a standardized PPA and a predictable purchase rate, along with private-sector financing, are essential elements of a successful PPA program. The 2003 reforms in the LOI process for SPPs seem to be positive. They should discourage lone speculators who do not have the experience or credit to support actual project development. This should improve the efficiency of SPP project development by prequalifying SPP developers and by eliminating indefinite hoarding of the LOI rights to the best hydroelectric sites.

Power purchase agreements

The principal features of the power purchase agreements are given in table 7-1. Table 7-2 chronicles the SPP tariffs from 1997-2003.

Table 7–1: Features of Sri Lanka SPP PPAs

Feature	Description of SPP feature
Basic provisions	
1. Parties	The contract is made directly between the state utility, CEB, and the SPP. As a body corporate, CEB waives the sovereign immunity it otherwise could assert against legal action.
2. Milestones	The SPP contract contains negotiable date milestones for (a) achievement of all necessary permits for land acquisition, construction and operation, and (b) achievement of commercial operation. The SPP is responsible for obtaining all permits.
3. Delivery of power	CEB must accept all power at the delivery point as long as operated pursuant to Good Utility Practices and the facility maintains its eligibility for SPP status by selling (not necessarily installing) no more than 10 MW, unless the CEB system is not able to accept power. The contract is only for the transaction in energy, not energy capacity.
4. Output guarantees	The SPP maintains control over the amount of energy sold, with the SPP designated as a "must run" facility, whereby CEB is obligated to take and pay for the energy tendered, unless there is an emergency in the CEB system. There are no consequential damages for which the SPP seller is liable, unless it diverts energy or heat to purposes other than sale of pledged output to CEB. If the facility is capable of generation, it must generate and deliver power to CEB. It may not divert power to other buyers. It may cease to generate only where there is a valid engineering reason for such interruption, and is obligated to provide at least 24 hours notice of interruption when possible.
5. Engineering warranties	Power must be delivered pursuant to IEC standards. The quality of the electric energy output delivered at the termination point is individually defined as to voltage, power rating, power factor, maximum line current and power, and frequency. Delivery voltage is 33 kV plus or minus 10%.

Sale elements

1. Power quantity commitment

    The output capabilities of the SPP are stated in the PPA. The SPP may sell no more than 10 MW of equivalent energy output under the contract. It is not prohibited for the SPP to install greater capacity than is sold to CEB.

2. Metering

    CEB owns and maintains the metering equipment. Either CEB or independent third-party calibration is allowed by contract, however, the contract does not specify how the parties choose from among these two alternatives. The meters are required to operate subject to IEC standards. Meters are tested annually and require accuracy within 2%. In the interim period, the SPP can request a test if it believes that the meters are not registering accurately, but regardless of the outcome, the SPP pays for such test.

    There is established a hierarchy of which set of multiple meters is employed to measure the energy sold during each billing period, cascading to secondary metering sources when the primary metering is not within accuracy parameters, and assuming that the secondary meters are operating accurately. If not accurate at the secondary level of metering, historic data from the prior year is utilized, adjusted by rainfall, stream flow, fuel consumption, heat rate, hours of operation, native self-use, and other factors, to estimate output. If this data is not available, data from the prior six months is used as an average proxy of the amount of output sold.

3. Net metering or exchange

    Not contemplated by the contract nor allowed by the program.

Risk allocation

1. Sovereign risk and financial assurance

    By contract, sovereign immunity is waived by CEB, as a body corporate, as a defense to suit.

2. Currency risk

    The tariff is paid in local rupees on a kWh-delivered basis. There is no indexation to foreign currencies. Therefore, borrowing in local currencies is necessary to protect against currency fluctuations affecting repayment options.

3. Commercial risk

    All commercial risk is absorbed by the SPP. The obligation to attempt to produce and deliver, and for the utility to take and pay for, energy, is absolute except for short justifiable interruptions on either side of the transaction. The term may be up to a 15-year term.

4. Regulatory risk and change of law

    Although there originally was a change of law clause covering regulatory and tax changes to facilitate a consequent adjustment of the price term, as suggested by the legal consultants, that clause was later not carried forward in the final PPA by the utility. Such risk is now borne by the SPP. The price paid for power is not based on any capacity payment. Only an energy component is paid, and this value fluctuates annually. Thus, there is no long-term certainty for the tariff, which impedes financing.

5. Excuse and force majeure

    Force majeure is provided for both acts of God and for more controllable acts. Force majeure is defined in a manner conventional for power sale agreements, including civil disturbance and failure of the sovereign to grant necessary permits. Failure to obtain necessary fossil fuel from a supplier for the SPP, or any other cause out of a party's control, is also deemed to be a force majeure event. The time limit for the maximum duration of a force majeure event is three years. After three years, if not cured, the other innocent party may elect to terminate after an additional notice of 90 days. This is at the most liberal allowance of the range of U.S. small power contracts surveyed by this author. This provides more flexibility to attract small power producers.

Transmission

1. Transmission and distribution obligations

    The SPP must deliver the power at its own cost to the delivery point, which is the line side of the isolator on the CEB grid, and pay for all interconnection and protective costs, as well as all interconnection costs up to the termination point. The title to energy passes at the metering point. CEB must use "best efforts" to take the power or to minimize any disruption given the "must run" status of the SPP. Since CEB is the only entity to whom the SPP may sell power, other than its host or otherwise allowed by license, there is no obligation of the utility to transmit power. Long delays and bottlenecks have been reported by one stakeholder, although it is not clear whether this is a persistent or isolated issue. CEB requires that it build the interconnection or an entity approved by CEB build the line to CEB design standards using materials purchased from CEB.

2. Interconnection arrangements

   Interconnection standards are governed by Interconnection Guideline G. 59/1 of the British Electricity Association. Either the utility can build and bill the SPP for the interconnection upgrades and equipment, or the SPP can construct the interconnection equipment pursuant to utility review and standards, and then dedicate such facilities to the utility. In either event, the SPP incurs the entire cost of the protective equipment. If upgrades, repairs or modifications are later required by the utility, the SPP must implement same at its own expense.

Tariff issues

1. Type of tariff

   The tariff is based on avoided cost principles. CEB defines its avoided costs as the maximum value of additional generation that it avoids. Even where Sri Lanka requires capacity and these projects effectively provide long-term capacity, and the PPAs are set up to provide "firm" power, CEB pays only an energy value for the power. The contract includes the marginal cost of fuel and variable O&M in this value. The marginal fuel cost is the cost of petroleum to CEB from the state Petroleum Corporation. Allowances of a total of 7.7% are made for station and transmission losses and CEB overheads.

   The tariff is the average of (a) the prospective energy avoided cost calculation and (b) the calculation utilized during the prior two years. Although initially recommended to calculate a three-year forward average of avoided cost, CEB elected to calculate a three-year backward average of avoided cost. Even for energy, this would have exceeded 6 cents per kWh in 2003, according to an independent consultant report in 2001. Because these tariffs would exceed those paid to IPPs, CEB elected not to pay full energy avoided cost to SPPs which are not dispatchable.

   The price for energy is seasonally adjusted; that is, it is reestablished annually by each December 1 for the ensuing calendar year. The "Dry Season" comprises February through April, and the "Wet Season" comprises May through January. These variations during each year are set forth in table 7–2.

2. Capacity obligations

   The SPPs were originally told that they would be paid a flat price of 6 cents per kWh, escalated during the term of the contract. This was later changed by CEB with payments to be set annually at the short-term avoided cost of CEB, with diesel fuel typically the fuel at the margin. CEB does not pay an avoided capacity component even for a long-term PPA that provides capacity value, even where the CEB system needs capacity. The energy component fluctuates annually, providing no long-term certainty for the tariff.

3. Fuel price hedging

The projects built under the SPP program to date in Sri Lanka are small hydroelectric projects. There is no fuel hedging for such projects.

4. Update mechanism

Energy prices are updated annually by December 1 to apply for the prospective year. The price is seasonally adjusted within each payment year. Energy prices may not decline below 90% of the price applying at the time of the execution of the contract. This was done to establish a floor underneath renewable energy prices so as to facilitate their financing.

5. Tariff penalties for nonperformance

Late payments bear interest at the prime rate then in effect.

Performance obligations

1. Operational obligations

The SPP must use its best efforts to deliver power. However, failure to deliver power for short periods, while justifying damages to the purchaser, does not rise to the level of a cause for termination. Provided in this contract are requirements for the SPP annually to forecast the amount of power to be produced and sold, with a minimum one month notice of planned outages, and the right of CEB to have access to and inspect the SPP facility.

2. Definitions of breach

Typical commercial definitions are employed. Failure to achieve milestones, failure to pay for 90 days, or bankruptcy of the SPP constitutes a breach. There are no express remedies provided for breach and no explicit penalties in this contract. There are no consequential damages. No deposits or other security are required of the independent producer. Breaches must be cured as soon as possible. A party has 60 days after notice to cure a breach without it constituting a default; or if it requires longer, such cure must be begun within 60 days and the cure accomplished within no more than two years.

3. Termination opportunities

Termination may not be made at the sole election of either party without cause, but may be made 30 days after default, which is defined in the agreement as an uncured breach that ripens into an event of default. Cause for termination includes only uncured default, uncured nonpayment, or uncured force majeure. The project lender gets an opportunity to cure any default. This is an important element for project finance in providing additional loan security to project lenders.

4. Guarantees of payment and performance
The Agreement contains no guarantees of any performance obligations.

5. Assignment and delegation
Other than to subsidiaries for the purposes of financing or to hold the project in a project company, the SPP may not assign or delegate its rights without the prior written consent of CEB, which may not be unreasonably withheld. A succession clause is included which has any successor of either party assume all duties. There is no restriction on assignment by CEB.

6. Dispute resolution
The parties first pledge to attempt to informally settle any dispute among themselves during a period of 30 days. If not settled, and the sum in dispute is less than SL Rs. 1 million, the parties may agree to appoint a single neutral party to resolve the dispute or may ask the government to appoint an expert in the field to resolve the dispute, in either case to resolve the dispute within an additional 60 days. If not then resolved within 90 days, either party may refer the dispute to arbitration under the Arbitration Act No. 11 of 1995.

Even though the price at which the CEB will purchase power was raised more than 75% in local currency between 2000 and 2002, this increase was offset by a 33% decline of the rupee against the dollar over the same period. This makes the net increase in the SPP power sale price (in dollars) equal to slightly more than 40%—still substantial. During the life of this program, the price for power has fluctuated between US$0.043 and US$0.06 per kWh, depending on the increase in SPP price relative to the depreciation of the Sri Lankan currency against the U.S. dollar.

Table 7–2: Annual Season SPP Tariff in Sri Lanka

Year	Dry season purchase price (in current terms)		Wet season purchase price (in current terms)	
	Rs. per kWh	$ per kWh	Rs. per kWh	$ per kWh
1997	3.380	0.068	2.890	0.058
1998	3.510	0.062	3.140	0.055
1999	3.220	0.047	2.740	0.040
2000	3.110	0.043	2.760	0.038
2001	4.200	0.051	4.000	0.048
2002* (early)	5.130	0.053	4.910	0.051
2002* (late)	5.500	0.057	5.650	0.059
2003	6.060	0.062	5.850	0.060

Source: Ceylon Electricity Board
* The tariff changed midway through 2002

Notes

1. Bandaranaike, Romesh Dias. 2000. Grid connected small hydro power in Sri Lanka: The experiences of private development. Paper presented at the International Conference on Accelerating Grid-Based Renewable Energy Power Generation for a Clezan Environment, Washington, DC. Most of these delivered shaft power for tea plantations. DC power was also provided for lighting at the plantations. These provided about 10 MW.

2. Bandaranaike, Grid connected small hydro power in Sri Lanka, 4.

3. Bandaranaike, Grid connected small hydro power in Sri Lanka, 4.

4. Bandaranaike, Grid connected small hydro power in Sri Lanka, 3.

8 Vietnam: Capital Markets and Renewable Energy in a People's Republic

Program Overview

Vietnam is a nation that is embarking on independent power development and has a shorter track record than some of the other Asian countries highlighted here. Vietnam has many elements in common with some of its Asian neighbors with regard to renewable energy and small power potential. Like Sri Lanka, it is a hydroelectric-based system. Like Indonesia and Thailand, it has local access petroleum and natural gas hydrocarbons to generate fossil fuel-based conventional energy. Like Indonesia, Sri Lanka, and India, the need for new investment in additional power generation capacity outstrips the available internal capital resources. Therefore, the attraction of additional private investment capital is very important.

Unlike Indonesia, Vietnam is not a fragmented island system of distinct grids. There is a central transmission spine running the length of the country that supports a national power grid throughout the country and reaches all 64 provinces, 96% of the districts within the provinces, 78% of the communes, and 69% of the households.[1]

Supply Resources and Demand

The most recent figures indicate that per capita electricity consumption is 390 kWh per year per person, still one of the lowest amounts in Asia. Demand for power is rising rapidly, and a high growth rate is expected to continue over the next 15 years, the end of the current forecast period. Between 1995 and 2000, total electricity sales increased 15% per annum on average. The load forecast approved in May 2003 by Electricity of Vietnam (EVN), the state utility, shows the following base-case load forecast actual and projected values at five-year intervals[2] (table 8–1):

Table 8–1: EVN 2003 Electric Load Forecast

Year	Gross Gen (Gwh)	Load Factor (%)		Peak MW
Actual				
1995	14,617	61.9		2,696
2000	26,595	62.09	4,890	
Projected				
2003	40,329	62.89	7,165	
2005	53,438	64.0		9,118
2010	96,126	68.44	16,033	
2015	141,260	69.0		23,370
2020	201,367	71.0		32,376

This load forecast reveals that the rate of demand growth is ending a decade of approximately 15% annual increases; the projection for 2005–2010 is expected to be 12.5% per annum, and, for each of the successive five-year periods, is projected to be 9% and 7.5%, respectively.

The projected demand is not the true gross demand. Rather, it reflects the system's ability to supply. With additional supply, additional demand would materialize. Most significant industrial and commercial establishments have their own on-site generators. Electricity demand peaks in the evening hours, as well as during the dry season (December–May), when the predominantly hydroelectric capacity of the EVN system has fewer supply resources to draw on.[3]

At the end of 2002, total installed and connected generation capacity was 8,749 MW, of which 8,450 MW was available capacity.[4] Of this, 47% was hydroelectric capacity, 25% gas turbines, 14% coal-fired Brayton-technology thermal plants, 6% oil-fired Brayton thermal plants, and 5% diesel facilities running on oil. During the rainy season in Vietnam, which runs from mid-June until mid-November, hydroelectric generation constitutes more than 60% of system generation, less during the dry season.[5]

EVN's Master Development Plan calls for the addition of about 12,000 MW of new capacity, owned by EVN and independent power producers (IPPs), between 2001 and 2010. Of this total, 2,400 MW will be small, medium, and large hydroelectric installations ranging in size from 1 MW to 300 MW. The plan contemplates 4,820 MW aggregate in 11 EVN-financed thermal plant additions of 160–750 MW each, 2,660 MW representing six IPP thermal power projects of 100–720 MW each, and 566 MW aggregate in seven IPP hydroelectric power plants of 65–100 MW each. In total for all technologies, this represents a target amount of 11,936 MW.

Vietnam has access to fossil fuels. The country produces oil and is constructing its first two refineries, scheduled for completion in 2005 and 2010. Vietnam has significant coal reserves in its northern provinces, which produce coal for export above and beyond domestic coal requirements. Coal extraction is estimated to be able to reach 15–20 million tons of coal by 2020. Vietnam also has large offshore natural gas resources. Crude oil production is estimated to reach 25–30 million tonnes per annum, and natural gas production could rise to 15–30 billion cubic meters per annum.

Institutional Differences in the Vietnamese System

What is different about renewable energy development in a communist state? Well, everything…and nothing. From an institutional and structural perspective, everything is different: Approvals, the role of the feasibility analysis, the state-connected role of financing, state sponsorship of developers, and, ultimately, the state source of lending, are all very different.

However, from the perspective of designing a program to efficiently attract international financial resources to the power sector of a communist country, nothing is different. For Vietnam to attract international risk capital, it must graft conventional legal structures, contracts, and enforcement provisions onto its institutional structure. The concepts of established, standardized, and enforceable power purchase agreements (PPAs), providing medium- or long-term avoided cost tariffs with financeable provisions, as well as a neutral regulatory structure with transparent and manageable approvals, must be grafted onto the system before capital will flow efficiently to any country.

The systemic institutional and legal differences in the Vietnamese system are significant. In Vietnam, government committees and agencies at both the federal level and in the 64 provinces institutionally make decisions that are traditionally made at arms length in noncommunist systems by multiple actors in the marketplace. These decisions involve electric energy investment, financing, and development. In essence, a set of interconnected and sometimes overlapping government actors function in the roles of—or control the functions of—the project developer, investor, feasibility analyst, constructor, banker, power seller, power purchaser, project owner, operator, and arbitrator. Banks are state seeded with capital and encouraged to make certain extensions of investment capital. Private development companies may be joint venture companies controlled by a majority interest of a state-owned agency.

Consequently, the normally pivotal role performed by conventional PPAs, finance commitments, and tariff provisions are different. For example, in Vietnam, power projects may be financed and constructed before there is an established tariff or executed PPA. The PPA and the tariff may be established well into project construction. Because the government bears all of the risk in its overlapping roles, it is assumed that projects will receive tariffs based on their costs of production, and the tariff and PPA may be adjusted during the project term. There are no competing actors or checks and balances. Ultimately, most of the risk of project development remains with the government.

In other countries, this would be unthinkable—the PPA is the foundation that must first be established for the power project development. In Vietnam, because of the overlapping of roles, and ultimate state authority over most facets of project development, as well as the experience of work among a small community of interested government-owned or supervised entities, there is a communality of purpose. Ultimately, the government is responsible for all sectors and bears all of the risk. The conventional checks and balances of the market are replaced by numerous committees that must approve the project.

How does this alter the nature of electric-sector development? Fundamentally. There is no conventionally functional market, as in the other countries surveyed here. A market economy in the power sector is still evolving. In a conventional market, risk is allocated by enforceable contract documents that allocate risk to actors who are best able to shoulder that risk in return for negotiated risk premiums. Risk is diversified and reallocated among those structures to absorb that risk in return for risk premiums provided to different stakeholders and actors.

In the socialist/communist model, government involvement with project developers, construction companies, lenders, utilities, and project operators gives the government control at every level, typically exercised through multiple committees. But these committee approvals can be duplicative, and there is no guarantee that project efficiency will result in any phase of project development. The ultimate price received by the small power provider (SPP) reflects its costs of construction and production. The system sets SPP power sale prices on a traditionally cost-plus basis. There are no market incentives to cut costs or to operate efficiently. Therefore, if the system makes a decision that a particular project is feasible and should proceed, it will be expected to achieve typical or average costs of operation. No entity has a financial incentive to make the project more efficient. This leaves the government bearing all of the risk. Risk is not allocated to the stakeholder that is most able to bear that risk. There is no ultimate diversification of risk. This model does not facilitate international capital flows.

In 2004, Vietnam began a process to identify the institutional impediments and needs for a viable SPP market. This is expected to lead to a series of regulatory modifications and system accommodations to make the SPP process more streamlined and the sector more transparent and attractive to international developers, lenders, and venture capital. Vietnam, by implementing its SPP program a few years later than the other countries discussed here, has the opportunity to analyze what has worked best in other systems and why, and it can benefit from the experience of these other Asian countries in designing and implementing its own SPP programs.

Program Design and Implementation

Between 1996–2004, Vietnam was engaged in the preparation of a new Electricity Law. As of now, it is in its 20th draft and is neither publicly available nor enacted.[6] When it is eventually enacted into law by the Socialist Republic of Vietnam, that law as now constituted will make several fundamental changes to the electric sector of the country. It is designed to enhance the efficiency and effectiveness of the power sector to adapt to its growing economy and to help achieve the delivery of electricity to the entire population of the nation. It will work to harmonize an often contradictory and inconsistent system of decrees and directives from different ministries and provide the regulatory framework to help transition the nation from the traditional government-managed electric structure toward a socialist market economy for electricity.

The Electricity Law promotes certain structural changes and reforms, but it does not go as far as other countries in Asia have gone. A series of roundtables and policy planning sessions have been held in conjunction with ongoing efforts to amend the Electricity Law. The Energy Sector Management Assistance Programme,[7] a joint effort of the United Nations Development Programme and the World Bank, has catalogued stakeholders' recommendations for additional elements to be added to the draft Electricity Law to make more fundamental reforms, including the following:

- Separate generation from transmission and distribution.

- Create a wholesale power pool.

- Allow independent wholesale power competition.

- Establish fair transfer prices for bulk wholesale power supply.

- Establish clear tariff principles to allow full system cost recovery and provide incentives and proper return on capital investment and expansion.

- State management of different functions of the electric sector—ownership of assets, policy, and regulation—should be assigned to different agencies, with assurance of clear lines of independent authority and no overlap between agencies.
- All policies need to be transparent.
- Establish a regulator that is independent from, rather than within or reporting to, existing government agencies.
- Gradually establish precedence for a regulator that is independent of other government agencies and actor who make decisions based on rules.
- Firmly embed in law the full rights to private-sector ownership and management of electric-sector assets.
- Segregate the fulfillment of the government's social obligations (rural electrification, road and infrastructure development in remote regions) from the government's budget, not from electric-system tariffs or prices.
- Institute competitive bidding for new supply of power.
- Simplify licensing arrangements for small systems and remote applications.
- Simplify tariff setting for small systems and remote applications and provide for long-term tariffs with periodic adjustments based on transparent principles.
- Ensure protection against the revocation of permits, except for causes articulated in transparent regulations and with rights to challenge revocation.

To accomplish its full objective to attract foreign investment and international capital flows to Vietnam's electric sector, lawmakers will need to be informed by these recommendations to make electricity reform as far-reaching and to adopt as many market mechanisms as possible. There is no specific treatment of renewable power or SPPs in the draft of the Electricity Law as it now exists. The draft law focuses on larger structural reforms, many of which are corollary to implementing a system that facilitates IPP or SPP development in Vietnam. Stakeholders are recommending that the draft incorporate additional provisions to further open the sector to private investment. Some of the recommendations address the simplification of permitting for small projects.

Ultimately, an SPP program will have to be developed through a specific program initiative in the context of ongoing electric-sector reform efforts. The SPP program planning and design in Vietnam began in 2000, but a formal program was stalled. Some SPP development progressed slowly and informally in the absence of a formal program. Vietnam initiated SPP program planning anew in 2004, with new consultants. The discussion in this chapter focuses on the current demand and supply balance and current resource potential and SPP commitments; however, because the SPP program is still in the development stage, it analyzes the PPA from 2000, which was not adopted but is the most recent standardized document.

Solicitation of SPP participation

The program in Vietnam is in the initial stages. A draft PPA was created by the original consultant and accompanied by a tariff design in 2000.[8] Since then, the final format of the PPA and the tariff have not been agreed to by the utility and government, and no formal SPP program has begun. In the interim, EVN has entertained IPP projects on a less structured basis.

In response to private-sector interest, EVN's decision was to allow IPPs to generate and sell directly to customers in designated industrial zones. The following types of private participation are available in principle:

- Private generation selling exclusively to EVN on a 20-year build, operate, and transfer basis

- Private generation selling exclusively to businesses in enterprise zones

- Private generation for a company's own consumption

- Private participation in distribution.

Size and resource limitations

To date, EVN is at the start of its IPP and SPP program and has not made a distinction between the two. It has committed to three IPPs, all of which are large fossil fuel–fired facilities. At a workshop in December 2002, EVN committed to an SPP program based on avoided cost principles,[9] although there is not yet an avoided cost tariff. With support from the World Bank, it began program planning and design anew in 2004.

The state utility envisions an affordable price for energy of about US$0.045 maximum during the dry season and US$0.025 during the wet season. This is limited because EVN does not want to pay more for energy at wholesale cost than it can charge for energy at state-approved retail rates, allowing the necessary retail markup of US$0.016–$0.018 for transmission charges. Therefore, although it is cognizant of the principle of avoided cost pricing, EVN feels limited by the pressure not to lose money under a system that is not operated to break even, but to advance the social policy of distributing electricity at rates that knowingly do not cover all revenue requirements. The EVN is committed to the principle of avoided cost payments, as long as they do not exceed its average retail tariff, less transmission expenses.[10] Avoided cost is a marginal rather than an average cost concept. In addition, EVN wants operational control of the plant, which could conflict with cost-recovery issues: The tariff is paid only on kWh sold, and failure to dispatch the plant would result in less revenue unless a split tariff were employed to assure that the capacity costs of the plant were repaid by a fixed, nonvariable tariff component.

Power authority and government role

Vietnam is a socialist country. This dictates that the economy—including the electric sector—be centrally planned. Therefore, factories are located by the central state planning authority in locations that take advantage of existing and planned electric generation and transmission capacity.[11] This fact distinguishes it significantly from the other Asian nations reviewed here, where geography or a lack of central planning causes industrial load centers to be located in disparate pockets of the country.

Enterprises are typically owned and operated by the state. The capital-allocation process is different for these companies because they do not account for profit and return in the same manner as publicly traded private companies. They depend on various levels of state approval for their access to capital and authorization for investment.

The approval process for projects in Vietnam is lengthy and involves multiple ministry approvals.[12] These include the submittal of SPP plans (sometimes) through a prefeasibility study, investment approval, establishment of a domestic or foreign investment activity, feasibility study approval, negotiation of the PPA and interconnection agreement, negotiation of the tariff, and construction approval. For an SPP project company, navigating the regulatory framework of the electricity sector is relatively complex and may include establishing a business entity, investment approval, and construction permit. The following steps are typically involved:

- Preparation of a prefeasibility study[13]

- Execution of a memorandum of understanding on the terms and conditions of the PPA

- Preparation of a feasibility study and environmental and social impact assessment for approval by the Evaluation Committee[14]

- Approval of the or memorandum of interest by the local development investment committee[15]

- For domestic investors, establishment of a business entity in the form of a limited liability company or shareholding by obtaining a Decision of Entity Establishment (operation license) from the chairman of the people's committee for private groups or from the prime minister or other relevant minister, and then registering the business activity with the Ministry of Planning and Investment or the Provincial Department of Planning and Investment

- For foreign entities, approval of the prime minister and issue of an investment license by the Ministry of Planning and Investment

- Execution of a power purchase agreement with EVN

- Obtaining a construction permit by submitting applications to the Department of Construction for approval to construct; to the Department of Lands and the People's Committee for approval of land use, clearance, and compensation; to the Department of Trade and Department of Customs for an import license for imported plant and equipment; to the EVN or people's committee for interconnection to the grid; to the Department of Labor for registration of staff; to the Department of Tax for issue of a tax code for value-added tax; to the Ministry of Finance for approval of accounting methods; and to the State Bank of Vietnam for registration if the company is eligible for the government's foreign exchange balancing guarantee.

Number and capacity of SPP interest and applications

Table 8-2 summarizes existing and proposed hydroelectric SPPs in the country that have made significant progress toward the necessary approvals. Although 3 of the 27 projects exceed 50 MW in capacity, most are small SPPs.

Table 8–2: Hydroelectric Project Status in Vietnam

Status/ Capacity Range (MW)	Projects (number)	Capacity (MW)	Generation (Gwh)
Operating 8–72	4	101.4	436.1
Under Construction 8–240	8	312.5	1,145.1
Proposed 5–18	15	152	639.9

A more liberal count of planned projects in the country increases the total megawatts planned. More than half of the 88 hydroelectric projects for which planning is under way are under 10 MW, and thus can be classed as renewable energy small power projects. Construction costs range from US$1,000 to over $2,000 per kW.[16] Approximately 67 proposed renewable energy SPP and IPP projects were planned in Vietnam as of the end of 2002.[17] Most were still in the development stage in 2005, with many lacking a signed PPA with EVN. All of these projects are sponsored by state-sponsored entities, people's committees, or communes, with the exception of one private 3.6 MW hydroelectric proposal. Some of the state projects may attempt to sell shares to private investors through joint stock companies.

Criteria for award

No criteria have been established for awards implemented by EVN because a formal SPP program is still in the development stage. In Vietnam, the feasibility study plays the role that stakeholders play in a market system. The feasibility study determines the project budget, and the tariff is then set to cover the costs of the project. There are a variety of ad hoc criteria applied by EVN. The project must have completed its prefeasibility study before contracting with EVN. Stakeholders are concerned that the largest state-affiliated construction companies will be able to determine the best projects and lock them up years in advance without a competitive process. There are multiple approvals at various levels of national and provincial government, which ultimately appear discretionary.

Award data

By 2003, only three large fossil fuel–fired IPPs had executed contracts with EVN. Vietnam entered its first large IPP projects with two international companies to construct large, combined-cycle, gas-fired machines of approximately 350 MW each. These projects involve indexation of payments. The contracts pay the developers approximately US$0.04 per kWh, which approximates EVN's avoided cost.[18] The average retail service tariff is US$0.056 per kWh.[19] All amounts paid to the smaller IPP and SPP projects are paid in local currency, with no indexation for foreign exchange. These small projects do not benefit from international agency guarantees.[20] The potential and operating renewable resource projects in Vietnam are displayed in table 8–3.

The potential renewable generation is significant. Vietnam has more than 2,200 rivers and streams longer than 10 km, which could yield an estimated 8.0–10.0 TWh from a total installed generating capacity of 18,000–20,000 MW.[21] Geothermal power capacity is estimated to be 200 MW, and perhaps as high as 400 MW; biomass cogeneration is estimated at 300 MW; wood and agricultural residues for the electric sector are estimated at 50 million tons per annum.[22] There also is the potential for solar and wind resource development.[23]

Table 8–3: Potential and Operating Renewable Energy Projects in Vietnam

Resource	Potential (mw)	Operating (mw)
Hydro Power	800–1400	110–115
Pico-Hydro	90–150	30–75
Isolated Mini grids	300–600	20
Grid-connected mini hydro	400–600	60
Off-grid solar PV systems	2	0.6
Biomass, bagasse, rice husks	250–400	50
Geothermal	50–200	0
Wind	TBD	0.2
Total	1802–3350	271–361

Size and type of technologies

The bulk of future SPP projects will be hydroelectric. Two gas-fired IPPs of about 350 MW each and one diesel oil IPP of about 300 MW have executed contracts. Most of the pending proposals awaiting decision are intermediate-sized hydroelectric projects ranging from 10 MW to 115 MW, although there are proposals for a 15 MW wind project and discussions of biogas from landfills. More than half of these proposals would be considered intermediate-sized IPP projects. According to EVN, almost half of the inquiries would be smaller (predominantly hydroelectric) projects of 30 MW or less. There are existing sugar mills, municipal waste, landfill gas, glass companies, shoe and apparel manufacturers, cement facilities, and rice processors that could be sites for cogeneration SPPs.[24]

Completion ratio and reasons for failure

The SPP program does not yet have enough experience to know the success-to-failure ratio. Interviews with stakeholders in Vietnam indicate that a lack of access to public or private capital is a major impediment. The sources of domestic capital are not diverse. Traditionally, the government owns all of the equity in energy enterprises. The bond market is underdeveloped and there is, as yet, no stock market. There is little or no access to long-term savings, and these institutions have no capacity to provide debt in the long maturities required for energy investments.

The lengthy process for project approval through the various state ministries was also mentioned as an impediment that could be streamlined. This uncertain approval process is exacerbated by the lack of a neutral, standardized PPA (EVN is negotiating PPAs in each circumstance and often after program construction has started) and the lack of a standardized SPP tariff. Because EVN is the only legal buyer of power, SPPs have little leverage in negotiation. Tariffs are set with regard to the costs invested in the project. Thus, all project costs are covered (measured against rules for typical civil engineering costs for hydropower projects), and no project is allowed an attractive risk premium or standardized price against which to maximize profit. Thus, the most cost-effective projects are not necessarily selected, and development does not experience cost-minimization incentives. There is no particular incentive or reward to complete projects within any project schedule. The lack of a standardized interconnection policy was also mentioned.

Process transparency

SPPs are now required to negotiate individually with EVN. The approval process for projects is long and complex. Stakeholders in Vietnam have asked for more SPP project transparency in the form of a standardized neutral PPA that incorporates a standardized SPP tariff, along with a standardized interconnection-operation policy. The utility is committed to such a process. Although several reforms have been made, the process is not transparent to all. There seem to be competitive advantages to some navigating the process, based on criteria other than neutral technical or financial measures. Even though the currency has been relatively stable compared to some other Southeast Asian nations (because the Vietnamese dong is subject to a managed float against the U.S. dollar),[25] this perceived lack of transparency has dissuaded many foreign developers and international lenders.

Stakeholder concerns

The state-owned developers have many of the same concerns that private SPP developers in other nations have about implementing a neutral standardized process:[26]

- The approval process is slow and cumbersome.

- The feasibility study approval requires a lender commitment, which is very difficult to obtain before having a PPA with EVN. The feasibility study process is complex and lacks uniform analysis methods across the many feasibility study (FS) services consultants.

- The interconnection procedure, which is not standardized and not dealt with in the PPA, is difficult.

- There is a lack of consistency in the PPA and in setting tariffs based on avoided cost principles.

- It is difficult to access capital from local banks that have high interest rates and often will not lend for a sufficiently long period; EVN receives a lower cost of capital than SPP developers.

- The EVN is unwilling to front-load the payment of the power revenue stream to correspond to the higher first costs of renewable power investments.

- The EVN has a tendency to pay lesser amounts for power when it believes that the SPP sponsor is receiving a higher-than-necessary return on investment, rather than enforcing an avoided cost or other consistent pricing principle.

- There is a lack of transparency in the negotiating process for PPAs and tariffs.

- The local state developers lack development experience.

- There is a lack of accurately identified and prioritized SPP projects (for various renewable technologies) in the Master Plan, so that the best projects are not identified and implemented where they are most feasible.

- Equal or controlled access to SPP project development opportunities and selection of developers is needed so that competition prevails.

- There is a lack of neutral dispute resolution.

- There is an ability to lock up future projects without developing them; a lack of competitive-bidding concepts, which would ensure cost-effectiveness; and a dominance of government-owned construction companies in the sector.

- There is no up-to-date list of and guide to the required permits and approvals.

- High transaction costs are required to develop small SPP projects.

- There are issues of enforceability of the PPA and the proclivity of EVN to change provisions of the PPA after it is signed.

- There is a lack of consideration of foreign exchange issues to attract international capital and a lack of escalation provisions for long-term contracts.

Lessons

Even in a socialist government structure, a neutral, standardized PPA that incorporates a standardized PPA tariff has great value and is necessary for successful SPP program implementation and administration. Without such a standardized program, resources are expended unnecessarily and inefficiently. The best projects are not identified and systematically developed, and risk is not allocated and spread among those most able to bear those risks.

Retail prices to final consumers are established and regulated by the government. Electricity tariffs for SPPs in Vietnam are based on the costs of production and distribution, which power companies submit to the Government Pricing Committee. Tariffs do not necessarily reflect the economic costs of supply, nor do they provide for a sufficient financial return. International experts, including the Asian Development Bank, estimate that Vietnam must charge 8.9 cents per kWh to attract investment and to recover capital in electricity system development programs.[27] Tariffs for SPP projects must be established based on neutral, internationally accepted avoided cost principles and ensured by enforceable, long-term PPAs.

It is in the interest of any nation that needs additional electric capacity to efficiently mobilize SPP and renewable energy resources. Experience in Vietnam indicates that it is not fundamentally different in a socialist/communist economy. When the objective is to mobilize private capital, a standardized PPA for SPPs is the most efficient, cost-effective, and transparent mechanism.

Power Purchase Agreements

In 2000, a consultant drafted a PPA for possible adoption by EVN. However, EVN has not yet adopted this draft PPA and likely will use a PPA of its own design when it does commit to a standardized SPP agreement. Like the agreements before it in Indonesia, Sri Lanka, and Thailand, the draft PPA is a tight and concise agreement that is appropriate for an SPP renewable energy project. It was drafted by the consultant after reviewing the Indonesian, Sri Lankan, and Thai PPAs, and it is expressly modeled on those—especially the Sri Lankan PPA.[28] However, it is different in important ways compared to the PPAs used in Indonesia, Sri Lanka, and Thailand. In particular, it leaves some terms less precisely defined and uses a "deemed energy" concept to pay the SPP for capacity, even when the utility does not take power. This was constructed to counteract EVN's practice of not committing to a firm take of SPP output. Even though it has not been implemented, this format contrasts with those in the other contracts mentioned here. The contracting provisions in the draft PPA for Vietnam, how they operate, and the principal features of the agreements are discussed in table 8-4.

Table 8-4: Features of the Not Implemented Year 2000 Vietnam SPP PPA

Feature	Description of SPP feature

Basic provisions

1. Parties

The contract is made directly between the SPP and the state utility, EVN. As structured, no lender rights are expressly recognized, as they are in the Indonesian PPA. Parties are allowed to sign the contract in two different languages simultaneously. No matter how proficient the translation, there will be significant differences and nuances that can change the interpretation. Ideally, there should be a single executed PPA for each project: The parties should execute only one contract, in either Vietnamese or English, typically at the election of the SPP so that it can utilize the language that facilitates project debt financing.

2. Milestones

A milestone for commercial operation is contained in the PPA, but its length of time is not specified. It is individually negotiated.

3. Delivery of power

EVN must purchase all power supplied by the SPP. No delivery requirement is imposed if there is a forced outage. EVN has indicated that it is willing to purchase all excess power if it has operational control over the SPP. This interface and control will need to be carefully structured during final negotiations on a standardized PPA.

4. Output guarantees

The PPA allows the utility purchaser not to accept or pay for power where SPP facility maintenance is inadequate, but it does not affect the quality of the energy. This allows the purchaser not to pay for deemed energy output, and could mask Transmission and distribution (T&D) problems. This could discourage lenders not to participate in this program.

5. Engineering warranties

The SPP must be operated pursuant to Prudent Utility Practices, which are conventionally defined, in a manner similar to the commonly employed concept of "Good Utility Practices."

Sale elements

1. Power quantity commitment

The agreement does not require the SPP to use best efforts to produce power (capacity) or EVN to accommodate and take power. There is no typical reciprocal obligation for the buy-sell transaction, where EVN must take, the SPP must produce and deliver. EVN can refuse to take power for any system-related reason. Otherwise, if it refused to take power, it must pay for deemed energy output.

2. Metering

The meters are maintained by EVN at the SPP's expense. The meters are calibrated at least every 12 months, with +/- 2% accuracy required. Secondary meters, installed at the expense of the SPP, are used to register quantity if the primary meters are not accurate; and if the secondary meters are not operable, estimation is done without any specific legal references for this estimation. So, this places the SPP at a disadvantage. There is no time limit on subsequent adjustment. There is a requirement that if one party thinks there is meter inaccuracy, the meters must be tested. The metering provision requires that both parties "shall" be present to break meter seals.

3. Net metering or exchange

There is no provision for net metering and no direct sale at retail is allowed the SPP.

Risk allocation

1. Sovereign risk and financial assurance

Nationalization or expropriation of the SPP assets by the government is deemed an event of default by EVN. However, the remedy for such is not clearly specified and could be difficult to enforce in any Vietnamese tribunal. After notice of default, the defaulting party has 60 days, plus an extension of another 30 days, to cure the default before it terminates the agreement.

2. Currency risk

SPPs would be paid in Vietnamese dong, so there would be no protection for currency fluctuations. The dong is subject to a fixed exchange and has been relatively stable.

3. Commercial risk

Commercial risk under the contract is borne by the SPP. The types of insurance required of the SPP are specified by contract without specifying the amount of coverage, any requirement to name the buyer as an additional insured, or any other requirements.

4. Regulatory risk and change of law

There is no provision on this risk.

5. Excuse and force majeure

"Force majeure" is defined as any third-party or extraneous action that interrupts performance. This would include failures of supplies, fuel, or T&D capacity.

"Forced outages" are defined only to include investigations, repairs, and replacement. Force majeure does not include failure to comply with EVN interconnection or grid standards or failure of a supplier to perform. Under the draft PPA, if something is wrong with the T&D system or repairs are necessary, then EVN pays for power it does not receive and gains no revenue because it cannot resell. However, if there is a force majeure event affecting EVN, there is no payment for this phantom energy. However, there is not a clear delineation between these two kinds of events in the draft PPA.

Transmission

1. Transmission and distribution obligations

Since there is no ability to wheel retail power or to make third-party sales, wheeling obligations of the utility to not arise. At the interconnection delivery point, the power becomes the property of EVN.

2. Interconnection arrangements

Interconnection is designed and constructed by EVN. These costs are billed to the SPP. SPPs are concerned about a lack of standardized interconnection procedure.

Tariff issues

1. Type of tariff

There is no standardized SPP tariff. The SPP power purchase price is negotiated on a case-by-case basis by EVN. EVN negotiates this tariff to attempt to not lose money on its retail resale of IPP and SPP power. Therefore, it subtracts from the average retail tariff its average transmission and distribution charges, yielding a residual value for the maximum SPP price. This methodology (a) utilizes average system cost concepts and (b) is limited by state-set retail tariffs for a system that does not earn revenues to cover its fully loaded costs. The avoided cost concept of SPP program design is predicated on marginal costs. Therefore, this tariff mechanism is not based on avoided costs at the present time in Vietnam.

2. Capacity obligations

The consultant in year 2000 recommended a dry and wet season tariff, with a $0.013 per kWh "minimum supply bonus" for SPPs that commit and deliver at least 70% of their capacity in a given month. This deemed energy concept is a concept that typically is associated with payment for capacity regardless of delivered energy. Every minor problem with EVN acceptance results in payment for output that cannot be taken. These types of tariffs and contracts typically are for large baseload fossil-fueled projects. Their adaptability to small renewable projects is not yet demonstrated in Vietnam.

3. Fuel price hedging

There is no fuel price hedging.

4. Update mechanism

There currently is no update mechanism. The retail tariff is expected by EVN to escalate to as much as $0.07 per kWh by about 2005. This could allow more flexibility for EVN to pay a higher cost for SPP power over time.

5. Tariff penalties for nonperformance

The tariff is not structured to encourage economical production and delivery of power on peak because the capacity payments are not loaded into the delivered power price. The contract implies that the project is dispatchable; however, this contract does not otherwise provide dispatch control to EVN. For small renewable projects, dispatch is not an ordinary operating paradigm.

Performance obligations

1. Operational obligations

The agreement does not require the SPP to use best efforts to produce power (or commit capacity) or EVN to accommodate and take power. There typically would be a reciprocal obligation—EVN must take, the SPP must produce and deliver. Where there are no penalties for nondelivery imposed on the SPP, there typically would be more flexibility for EVN acceptance. EVN can refuse to take power for any system-related reason. Otherwise, if it refused to take power, it must pay for deemed energy output.

2. Definitions of breach

Default occurs if permits cannot be obtained by the SPP. This failure, or an improper assignment or failure to carry insurance, could result in a default whether or not it would be deemed a material breach otherwise. Typically, only for *material* breaches are damages (but not default and cancellation) the appropriate remedy. Cancellation is not a particularly effective remedy for the SPP under certain default scenarios because there is no allowed net metering or other retail or wholesale power sale opportunity.

3. Termination opportunities
> The draft contract provides that even if it is the SPP that terminates the agreement because of a default by EVN, that EVN has purchase rights to the SPP facility at fair market value. This option becomes mandatory under certain termination conditions. This allows a one-sided buy-out provision. It is not specified whether the buy-out price at the time of termination is calculated to reflect the fact that the project is in default and therefore no longer viable. EVN is provided an opportunity to extend the term of the agreement for a contracted period of years at the specified termination of the agreement.

4. Guarantees of payment and performance
> There are no sovereign or other guarantees of performance by the utility. In a socialist economy, both the seller and purchaser of power are state entities.

5. Assignment and delegation
> Any assignment requires the prior written consent of the other party, which shall not be unreasonably withheld. Without consent the SPP can assign to an affiliate or for the purposes of financing the facility.

6. Dispute resolution
> If a dispute ensues, the parties shall try to settle the dispute informally for 30 days. If not resolved, the dispute is submitted to arbitration. The place of arbitration and the rules under which resolution is pursued are left blank for the parties to complete. Either party has the ability to cut off the other party's court rights by making a unilateral referral to the arbitrator.

Although it is patterned on the Sri Lankan PPA, the 2000 draft consultant PPA employs a different tariff concept than the PPAs employed in Indonesia, Sri Lanka, and Thailand.[29] These three countries employed a PPA adapted from the U.S. experience under the Public Utility Regulatory Policies Act. The Vietnamese draft from 2000 employs the concept of deemed energy.[30] This concept typically is not employed in SPP PPAs, to create a simpler format. Such a two-level or "split" PPA tariff is typical of larger fossil fuel–fired baseload facilities, and it is used less often for intermittent, smaller renewable energy facilities. This distinction is, in part, a reflection of EVN's desire to control the operation of IPPs in the country as a condition of power purchase, and a function of the consultant's recommendations. No final decisions have been made on the draft PPA and tariff design, which will be revisited in 2005-2006 as part of the reinvigorated SPP planning process.

Notes

1. Vietnam. Ministry of Industry. 2003. *Master plan study on electric power development in Viet Nam, summary report.* Hanoi, Vietnam: Ministry of Industry. This does not mean than 69% of households actually subscribe as customers and consume electricity, but that 69% of households have physical access to electricity.

2. Vietnam Institute of Energy. 2003. *Master plan study on electric power development in Vietnam: Period 2001–2010 and perspective up to 2020, base case load forecast.* Report prepared for the Ministry of Energy, Electricity of Vietnam (EVN). This forecast does not account for price-elasticity-driven increases in demand, which could be another dynamic factor in increasing demand for electricity. This could increase demand in ways that are not accounted for in recent forecasts for the sector.

3. Worley International. 2000. *Renewable energy small power purchase agreement.* Draft final report for Electricity of Vietnam. Auckland, New Zealand: Worley International.

4. Ministry of Industry. *Master plan study on electric power development in Viet Nam, summary report.*

5. Ibid.

6. The process of electric-sector reform through a change of laws and regulations in Vietnam is supported by the United Nations Development Programme and the World Bank.

7. World Bank, Energy Sector Management Assistance Programme. 2002. *The electricity law for Vietnam: Status and policy issues.* ESP259/02.

8. Worley International, *Renewable energy small power purchase agreement.*

9. Vietnam. Ministry of Industry. 2002. Workshop on SPP PPA, December 20. Hanoi, Vietnam: Ministry of Industry.
Professor Ferrey was present as a participant at this workshop.

10. Personal interview with EVN. Hanoi. December 2002. (Procedure was to assure anonymity).

11. Personal interviews with EVN, Vietnam Ministry of Industry, and World Bank officials, December 2002.

12. Worley International, *Renewable energy small power purchase agreement,* sec. 2. Before a project goes forward, the producer is required to obtain a prefeasibility study and approval, memorandum of understanding with EVN as to power purchase (which is now negotiated on an ad hoc basis), a feasibility study with approval and "stamps" from various ministries, establishment of a business entity for the project, a construction permit, and special approval for foreign investment.

13. For developments that are on the government's list of preferred projects, it is not necessary to prepare a prefeasibility study.

14. An evaluation committee is set up consisting of people from relevant ministries and government agencies to review and approve the development. For rural electrification projects, the reports are submitted to the people's committee or the ministry in charge. They will, in turn, pass the reports on to Evaluation Committee, and the approval process will apply.

15. Vietnam. Decree No. 52/1999/ND-CP of July 8, 1999. Hanoi. This decree classifies domestic investment projects into three categories: (1) projects with development costs greater than VND 400 billion (approximately US$26 million); (2) projects of VND 30–400 billion (US$2–26 million); and (3) projects less than VND 30 billion (US$2 million).

16. Price estimates are from various sources, including off- and on-grid installations.

17. Project list of EVN (2002). These projects are proposed by 13 state enterprises, municipalities, or communes, plus one private developer.

18. Personal interview with EVN, 2002. British Petroleum and Electricité de France are venture partners in these two projects. The price escalates annually for natural gas cost increases of about 2.5% annually. One existing diesel IPP project of about 375 MW, which receives a purchase price of about US$0.056, already exists.

19. Personal interviews with EVN, Vietnam Ministry of Industry, and World Bank officials, December 2002. The rates were raised in October 2002. The tariff is expected to continue to be raised in successive years. There is cross-subsidy in the tariff: Large users cross-subsidize smaller users, and rural consumers benefit from cross-subsidy as part of the tariff policy of the government.

20. Personal interviews with EVN, Vietnam Ministry of Industry, and World Bank officials, December 2002.

21. Economic Consulting Associates, Robert Vernstrom Associates. 2003. *EVN tariffs: Interim report.* (private consultant report by individual consultant) Bangkok, Thailand.

22. Ministry of Industry, *Master plan study on electric power development in Viet Nam, summary report.*

23. PROACT International, Inc. 2002. *Renewable energy master plan in the northern part of the Socialist Republic of Vietnam.*

24. One of the sugar mills has a two-year agreement at about US$0.037 per kWh to sell excess energy to EVN (personal interviews with EVN, Vietnam Ministry of Industry, and World Bank officials, December 2002).

25. The Vietnamese dong depreciates about 5% per year against the U.S. dollar in a managed float (personal interviews with EVN, Vietnam Ministry of Industry, and World Bank officials, December 2002).

26. Vietnam. Ministry of Industry. 2002. Workshop on SPP PPA, December 20. Hanoi, Vietnam: Ministry of Industry. Professor Ferrey was present as a participant at this workshop.27. U.S. Department of Commerce, National Trade Data Bank. 1999. *Vietnam: Leading sectors for U.S. export and investment.*
http://tradeport.org/ts/countries/vietnam/sectors.html (accessed September 3, 1999).

28. Worley International, *Renewable energy small power purchase agreement*, sec. 4.2.

29. Worley International, *Renewable energy small power purchase agreement*, 4.9.

30. Worley International, *Renewable energy small power purchase agreement*, appendix A.

9 Lessons Learned in Asia for a Successful SPP Program: Template and Techniques

Common Threads and Differences

The five Asian nations surveyed here use different forms of government and have different predominant fuel sources in their generation base (hydroelectric, coal, gas, and oil). Some of the national electric systems have an integrated high-voltage transmission system, whereas others have a disintegrated or *island* system. The following are the important similarities between these programs:

- All are in need of long-term increases in power generation capacity (although Thailand has a short-term current surplus).
- All have a mix of small-scale renewable energy options that could be developed.
- The installed generation base of each system is either small or disaggregated by island systems or state subsystems.
- Each country is being approached by private developers that seek to develop renewable energy small power producer (SPP) projects.
- Each system employs, either deliberately or informally, a standardized power purchase agreement (PPA), although it is not necessarily a neutral and consensual document in every case.
- Although avoided cost concepts for establishing the SPP tariff are recognized in each nation, they are not employed in all cases, or they are not employed to pay the long-term avoided cost of capacity when it is provided under the PPA.

There are some interesting common elements of tariff design. Most of the programs have not elected to index their tariffs to foreign exchange (Thailand's capacity payment is an exception). In some Indian states, the tariff is indexed to inflation to retain value in international currency amid local inflation. Other countries unilaterally review and may adjust the tariff at a prescribed time. This periodic adjustment does not provide long-term predictability for project finance or equity investment. However, the adjustment period or adjustment base varies in each country. Sri Lanka, for example, employs a three-year rolling average to attempt to smooth out volatility in imported oil prices, which set the marginal energy price. Other programs have implicitly capped their SPP payments below the actual avoided cost so as not to exceed the power purchase rate paid to larger independent power producers (IPPs).

As long as local lending is employed at the time of acquisition of any foreign component of the SPP, indexation or hedging against foreign currency exchange is not critical. Hedging becomes important only when international commercial financing is used, when repayment is owed to foreign equipment suppliers over time in foreign currency, or when a significant imported fuel component is required. With small and renewable projects, more local financing is used, and there is no foreign fuel component. Therefore, SPP programs can avoid some of the foreign exchange challenges that confront larger fossil fuel–fired IPPs in these same nations.

Although these SPP programs are designed to equalize some of the monopsony power that the state utility can exercise in its implementation, the utilities have often been reluctant to create the level playing field that the U.S. system for SPPs was designed to create under the Public Utility Regulatory Policies Act (PURPA). The experience in several nations indicates that when the utility is the only legal buyer of SPP power, the overwhelming advantage in bargaining power that the utility enjoys must be mitigated and balanced by a carefully designed and faithfully administered SPP program. Even when neutral rules suggested by consultants have been adopted in principle, they have sometimes been compromised in their application.

Key Issues in Renewable SPP Program Design

Several important lessons for SPP program design and policy have been revealed by the analysis of these programs.

Legal infrastructure

Dispute adjudication. A framework for structured SPP project development is necessary. SPPs do not spring fully born from the existing electric-sector environment. There must be a system of law, regulation, and utility interface that facilitates orderly SPP development. For example, stakeholders have reported difficulty in accessing neutral court adjudications in Indonesia for disputes involving both renewable and conventional IPP projects. The Indonesian SPP program was never successfully implemented, and access to commercial international capital in that country has been truncated.

Cost-based principles. Such a structured program operates best when utility operations are organized around and operate pursuant to the principles of cost-covering design and collection of tariffs. When a utility is self-sustaining financially and organized as a self-sufficient body corporate, its role is not threatened by SPPs, renewable energy projects, or wholesale competition. If it is not organized around these sound principles, such innovations can threaten the revenue base of the utility.

SPP project enhancements. Undercutting the value of the tariff to the SPP over time or the principles of the PPA causes SPPs to seek to sell power output to third-party buyers to realize the full market value of the power. In all programs, the first customers that are lost to the utility system when retail competition is introduced are those that have the most attractive load profiles and are least expensive to serve. Cream skimming the most attractive customers by private power retailers occurs whenever competitive supply is sanctioned in any system. Those most likely to exit the system are those who pay a tariff that is above their marginal cost of supply. Therefore, if the retail tariff structure in a country is not based on reasonable cost-based and revenue-recovery principles, any form of competition, whether it is net metering, energy banking, or retail sales, will

187

cause loads that are cross-subsidizing the system to be the first to use and benefit from these innovations. This can further erode system revenues. The system as a whole needs to be on sound economic, financial, and legal footing to make the transition to innovative legal and transactional structures for SPP development.

Scale of projects. The eligible maximum SPP size should be scaled to system capacity so that the program applies to smaller projects that will not significantly impact system capacity. Each of the programs surveyed limits each eligible PPA to no more than 0.5% of installed system capacity.

Allocation of risks. A variety of commercial, sovereign, currency, and regulatory risks are implicitly or explicitly allocated in the PPA. A basic risk is the way the PPA allocates the risk between the buyer and seller for nonperformance. There is great diversity in the way the six systems allocate this risk of nonperformance between seller and buyer. Some PPAs adjust the price paid for power to reflect capacity delivered, thus equalizing any disparities through price mechanisms. The Thai program reduces the SPP capacity payment when the SPP does not deliver, but it has no equivalent sanction against the Energy Generating Authority of Thailand for failure to take power. Other programs employ reciprocal legal obligations. The Indonesian and Sri Lankan PPAs, as they were originally designed, required both a firm SPP and the best efforts of the utility buyer to deliver and take power. However, the Indonesian State Electricity Corporation later unilaterally altered the Indonesian PPA to allow the utility to refuse to take power and not pay for this deficiency. The Andhra Pradesh PPA makes no payment for the capacity value of the power, but requires the buyer to accept all power delivered. The ability of the SPP to wheel power has been made economically prohibitive. By contrast, Tamil Nadu facilitates SPP power wheeling, but the SPP is required to back down power if the utility cannot use the power. Nonetheless, Tamil Nadu provides a market alternative when it does not accept power. In both Indian states, there is no firm delivery obligation on the SPP—it delivers on a nonfirm basis. In contrast, even though the PPA obligates neither party to sell or buy power, the Vietnamese program design requires Electricity of Vietnam to pay for "deemed energy," even when it does not accept SPP energy. The Vietnamese program design shifts the risk for nonperformance to the utility.

Interconnection. Utilities must interconnect with SPPs subject to a straightforward procedure to accomplish this without significant transaction costs or interconnection risk.

Financial catalysts. International funding, in the form of loans, Global Environmental Facility (GEF) funding, and Prototype Carbon Fund (PCF) payments for carbon credits, has been used as a catalyst for SPP development.

Solicitation of participation and competition

Transparent process. A transparent process is required to build investor, developer, and lender confidence. The program in Thailand operates by declaring in advance the amount of SPP and IPP resources that are being sought and then allocating entitlements and subsidies in order of the most preferred projects and the least required subsidy for renewable projects. This is an objective and transparent program rubric. It maximizes the available subsidy resources and uses the market to identify the most viable projects that need the least government subsidies.

Open offers. An SPP program can be initiated and sustained by an open offer to execute PPAs or by an ordered and time-limited solicitation process. Thailand and Indonesia employ the latter approach to control the amount and selection criteria for SPP project development. The Indian states surveyed and Sri Lanka have an open standing offer to purchase, although Sri Lanka is implementing more thresholds and control in allowing developers to freely accept a letter of intent (LOI) to vest development rights at a specific site. The former countries were concerned that they might have sufficient IPP power for the short term, and thus they sought to control the quantity of additional SPP power committed. An open offer allows a constant rolling development of SPPs, much like the original PURPA design in the United States.

Controlled solicitation. An ordered solicitation can inject competitive bidding, which, if correctly administered, results in bid-price reduction and competition for the best projects and sites. This has been well demonstrated in the Thai experience.

Milestones and bid security. There has been an issue in Sri Lanka and Indonesia with companies procuring LOIs or winning a solicitations without sufficient experience or resources to actually develop an SPP project at a site. This is particularly a problem with limited hydroelectric sites (the Thai program has not developed many hydroelectric sites). The developer's objective is to control either fixed hydroelectric sites or SPP PPA entitlements, and, in many instances, to sell that legal entitlement, once procured, to a sufficiently capitalized and experienced third-party developer. This process does not promote the most efficient and direct SPP development, but instead layers additional transaction costs and parties into the SPP development cycle. The Thai program requires a bid security deposit of 500 baht per kW (US$12 per kW) of capacity. In 2003, Sri Lanka placed a six-month limit on the validity of LOIs granted for SPP projects and required bid security bonds of Sri Lankan (SL) Rs 2,000 per kW (US$20 per kW). This does not eliminate the hoarding and selling of SPP entitlements, which is particularly vexing when hydroelectric sites are involved, but it does cause speculators to commit a forfeitable deposit and to deal quickly before the LOI terminates for lack of progress on a project. Although bid security deposits, limits on the assignment of rights, and time limits have a constructive role to play, they must be used intelligently so as not to truncate the SPP market. The Thai program also requires a deposit of 100 baht per kW (US$2.50 per kW) of applicants and more for larger sources.

Competitive solicitations. Organizing standardized SPP power capacity solicitations under a uniform set of rules provides greater program competition and uniformity of process than entertaining ad hoc applications. A competitive process allows the utility to compare all SPPs simultaneously against a standard set of criteria. In such a format, the utility retains control of the option to accept or decline SPP offers.

The basic PURPA design in the United States does not provide a competitive process, although some states have grafted a competitive process onto their PURPA mechanisms.[1] An organized solicitation allows a competitive process to drive down the bid price for SPP subsidies that may be provided, as demonstrated by the Thai SPP system. When the system needs additional capacity, an open offer attracts the most SPPs acceptances. When the system does not have an immediate need for capacity, a controlled solicitation of SPP bids provides the most control as a successful management practice. All of the states seem to recognize these appropriate practices.

SPP power sale enhancements

Avoided cost principles. The state utility has a monopsony on the purchase of wholesale power in most of the electric sectors of the nations of the world. This single-buyer model becomes significant with the partial deregulation of wholesale power supply. Although IPPs and SPPs are encouraged to supply power, they are still dependent on a single state buyer to both purchase and transmit their power. Many of these programs did not introduce a mature utility regulatory authority before they introduced an SPP power program. Therefore, it is absolutely essential that the utility operate in its role as purchaser and transmission entity subject to objective PPA and tariff principles. Such tariff principles include a tariff based on avoided cost principles. To efficiently promote renewable SPPs, either (1) a program for the purchase of all power at its full value to the wholesale system or (2) the introduction of some combination of third-party retail sales, net metering–energy banking, or third-party wheeling can be implemented to ensure SPP development. As state utility systems consider privatization, allowing retail competition can ameliorate the power of a single state utility purchaser and transmission provider. The Indian states have different policies on third-party power sales and net metering.

Renewable set-aside. The Indonesian and Thai program models offer an interesting approach. They set aside a certain entitlement (capacity) for SPP or renewable SPP projects. This allows the buyer to competitively evaluate the possible SPP proposals side by side. This allows a simultaneous decision based on comparability of proposals. This has also been done in the United States in states that have adopted competitive-bidding programs as an innovation of their administration of the PURPA program.[2] A 21st-century variant, to set aside an allotment for small renewable power, has been adopted in eighteen U.S. states that have moved to retail competition. In these eighteen states, a renewable energy portfolio standard requires a minimum percentage of the power sold by each retail seller to be derived from renewable power.[3] This forces retail suppliers, whether they are utility or competitive sellers, to obtain renewable power in the wholesale market. Tradable "green" credits can be used to inject liquidity into these markets.

Third-party sales. None of the five Asian programs currently allows direct third-party retail sales of power by the SPP (except in limited industrial estate areas). However, other states in India do allow direct retail sales, and the two states surveyed here allowed such sales by SPPs in the past before disallowing them. Thailand, which led the initial development of Asian PPAs, is considering moving toward an open market for third-party retail sales. Therefore, experience in other places with direct retail sales or net metering is particularly relevant.

Net metering and energy banking. A recent innovation in energy banking in three-quarters of U.S. states is known as "net metering." Net metering provides an incentive for SPPs by allowing them to "exchange" the power they produce and sell the utility during a billing period (typically a month) for power that they or their affiliates take from the utility. This buying and selling is netted at the end of the billing period. Several of the Asian countries have adopted energy-banking variants of this concept. Because the fully loaded retail price at which a utility sells power to the SPP is typically much greater than the wholesale avoided cost price at which the utility purchases power, net metering allows the SPP to sell an amount of power output at a price that exceeds the utility's avoided cost. This provides a significant subsidy to the SPP operation through the net metering or banking exchange.

Inventory of net metering. Most net metering applies exclusively to renewable generation and, in more limited situations, to fuel cells or cogeneration or to both. In the United States, net metering, implemented at the state level, is now regarded as the most significant incentive for promoting renewable energy and distributed generation.[4]

Tariff design

Renewable premiums. In systems that are experiencing current and projected shortages of grid-connected power resources, temporary payment of more than the avoided cost of renewable energy and small power development is one method of providing incentives for immediate power resources, and it can reflect the premium short-term value of additional power. However, there is risk in deviating from the avoided cost concept.

If the utility deviates from avoided cost principles at certain times, there is nothing to prevent it from deviating from them at other times. This may cloud the transparency of the process.

Tariff floors and subsidies. In many systems, additional subsidies are necessary to assist higher-cost renewable energy and smaller SPP projects. This is evident in some systems for which it is difficult to fully subscribe SPP program potential. This subsidy can be accomplished by explicit subsidies, as in the Thai system. Alternatively, there can be a tariff structure, as in Indonesia and Sri Lanka, that supports renewable power projects by placing a floor under the energy component of renewable power sales or that reflects a premium for the diversity value of renewable fuels.

Price bidding. Bidding can be employed strategically to minimize the ultimate system cost of renewable power resource development, as with the controlled bidding program for renewable subsidies in the Thai system. Indonesia developed a bidding program for SPP PPA entitlement, but it did not involve individually differentiated subsidies.

Renewable exemptions. A variety of national and state tax policies (such as direct production subsidies or credits, import duty exemptions, or income tax holidays or moratoria) can target the development of renewable SPPs and reduce higher effective construction costs. However, such subsidies typically apply to equipment for both SPPs and IPPs.

Tariff incentives. It is a successful practice to provide market incentives through the PPA structure. A PPA in which the incentive for SPP operational compliance is embedded in the energy (kWh) payment, rather than split into a fixed capacity payment plus an energy payment, maximizes the economic incentive for the SPP to deliver power at peak times. When power is not supplied, 100% of the potential revenue stream—embodying both energy and capacity if a capacity payment is included—is lost to the SPP. The alternative is to include legal sanctions in the PPA for nondelivery or other failure to perform. This would involve complicated decisions about the reason for failure to deliver and, ultimately, recourse to a reliable, prompt, and neutral arbitration process or court resolution. In many of the countries with smaller renewable power programs, it is more transparent and straightforward to embody these incentives in the tariff operation rather than embed legal sanctions in the PPA. Properly structured incentives can work more efficiently than penalties. That proper structure is a specialized exercise.

Capacity tariffs. The SPP PPA tariffs in the Indonesian, Thai, and Sri Lankan programs were designed to include capacity payments in the tariff payment for each kWh delivered, paid only if the SPP delivers power. This was designed to provide the maximum incentive to the SPP for dedicated performance and delivery of power while not invoking any coercive penalties against the SPP for failure to perform to a set standard. This type of simplified incentive is appropriate for renewable SPPs, for which the marginal cost of operation is relatively low compared to the initial capital investment. During the implementation of these tariff levels, Sri Lanka altered the tariff design to eliminate the capacity component for both short- and long-term contracts.

Standardized PPA. All programs employ standardized PPAs, and most employ either an avoided cost–based tariff or avoided cost principles with a cap on the tariff. All allow some form of long-term firm contract commitment.

Recommended Best Practices and Program Template

It is clear, in an era of robustly increasing demand for electric service, supply, and reliability in Asia and amid tight capital availability, that renewable energy projects and private capacity additions will have a premium value in most nations. Renewable projects avoid vulnerability to conventional fuel unavailability and price fluctuations, and they continue to draw from a naturally renewed and essentially costless energy input to electric generation. The renewable option is available in abundance in Asia. Every nation in the world has enough solar energy falling on its roads and buildings to supply 100% of its electric requirements.

Lessons Learned in Asia for a Successful SPP Program: Template and Techniques

The challenge is to develop these resources most cost-efficiently. Private SPP investments in electric-sector-generating assets accomplishes the needed and planned electric sector development without obligating additional state resources. The recent trend of increasing private SPP development will accelerate during the 21st century.

As some of these SPP programs have proved themselves, the project development environment has changed significantly. International investors have become more cautious. The general economic slowdown, coupled with the negative performance of international investments in IPP and transmission company systems, has made many international developers hesitant to assume as much risk with new private IPP or SPP investments as in past investments. Many have liquidated their international power investments. In addition, with nonperforming generating assets, wholesale power markets in many developed and developing nations have become illiquid. Equity investment and debt financing have become more problematic and conservative. Disagreements between IPP developers and utilities over PPA enforcement in several Asian and other nations have compromised investor confidence. Even with SPP success, the context and environment of private international investment has shifted.

To continue the trend of SPP project development, nations and utilities will have to be even smarter in the future. The playing field will need to be level, the PPA will have to be sufficiently neutral and legally functional, and the tariff will need to provide fair incentives for the risk and term undertaken. Haphazardly designed or implemented SPP programs will not succeed in the new environment. Nations and multilateral agencies will need to understand what works, what creates competitive advantages, and what leverages the maximum program resources most efficiently. Success in this environment will require both a desire to make a program work and the resources to be savvy about how to design the program and the PPA. In this sense, this book seeks to embody SPP program experience to date and to provide guidance.

These five Asian nations are pioneers in creating SPP programs to promote new renewable energy development. Each program alone has value in demonstrating and implementing the SPP programs embodied in enforceable PPAs. Cumulatively, however, they have provided an energy laboratory during the past decade that has left an inventory of more and less successful program techniques. Comparatively, they illustrate that certain techniques are the tools of choice as these and other nations craft renewable SPP programs as a significant element of their capacity plans.

The Asian SPP programs that are designed to foster renewable energy development suggest a range of successful practices for PPA development in these contexts. Table 9–1 highlights some of the successful practices that have been implemented in five programs in Asia.

What do this experience and changing power market trends mean for the future of SPP programs in Asia? Those nations that use the lessons of existing SPP programs will best use available resources and maximize SPP development. Regardless of one's perspective, in the 21st century, after 20 years of SPP program experience worldwide and 10 years of SPP programs in Asia, an international expectation has evolved for any PPA to be financeable.

Certain relationships, provisions, warranties, and sanctions are expected to be incorporated into any PPA if it is to provide the credit support for conventional debt financing and attract equity participation. There is experience in Asia with neutral PPAs that have achieved these requirements, as well as with unilaterally altered PPAs that became untenable to support an SPP program.

Although there are many ways to operate an SPP program, certain key elements of the PPA cannot be omitted or significantly changed without compromising the integrity of the program. The most successful future SPP programs will assimilate international PPA requisites as a base and layer on top of this base innovative, country-specific incentives, competitive embellishments, and risk-allocation techniques. To strengthen and build transparent and accepted SPP programs in this changing environment, the following techniques are relevant:

- **Tuning up the PPA.** A correctly structured PPA is the legal foundation of the program. Countries should perform a fresh examination of their existing PPAs to harmonize them with the lessons learned to date and the changing investment environment. This is a very quick exercise that can be performed by a legal consultant or a working group that reflects the input of all of the various stakeholders. Ultimately, a successful long-term program must operate pursuant to transparent

Table 9–1: Successful Management Design and Practices

Successful design and management practice features	Thailand	Indonesia	Sri Lanka	India: Andhra Pradesh	India: Tamil Nadu
PPA size <0.5% of system capacity	Yes	Yes	Yes	Yes	Yes
Open offer if need capacity	n.a.	No, but very large solicitation	Yes	Yes	Yes
Controlled solicitation if surplus capacity	Yes	n.a.	n.a.	n.a.	n.a.
Milestones on development time afforded SPP	n.a.	Yes	Yes	Yes, if NEDCAP financial guarantees	n.a.
Bid security deposit by SPP	$12 per kW	n.a.	$20 per kW	n.a.	n.a.
How renewable technologies are encouraged	Competitive award of subsidy	Hierarchy of renewable SPP preference; floor price on renewable power	Floor price on renewable power	Tariff differentiated for base load and intermittent renewable SPPs	None
Competitive solicitation	Yes	Yes	No	No	No
Standardized PPA	Yes	Yes	Yes	Yes	No, under development
Long-term firm PPAs	Yes	Yes	Yes	Yes	Yes
Avoided cost based tariff	Yes	Yes	Yes	Yes	Yes
Capacity payment for long-term power	Yes	Yes	No	No	No
Allocation of performance risk between seller and buyer	Alteration of capacity payment; utility can refuse delivery	Neutral; originally mutual best efforts	Neutral; mutual best efforts	Non-firm, but utility must accept all power	Non-firm, but utility can refuse delivery
Capacity payment adjustment if seller does not deliver power	Yes	No, capacity payments in peak rate	n.a.	n.a.	n.a.
SPP unit dispatchable	Yes, if firm capacity PPA: 80% minimum annual output purchase obligation changed	No, as PPA originally conceived; dispatchable without limitations after PPA	No	No	No
Wheeling, net metering, or energy banking	Energy banking	Wheeling	n.a.	Energy banking, wheeling	Energy banking, wheeling

n.a. Not applicable.

and objectively predictable PPA principles. These must include a PPA that provides a neutral control over the obligations and behavior of both parties and a tariff that reflects the avoided cost principles of establishing the value of power. No SPP program can operate at its maximum potential without these elements.

- **Utilizing competitive market design.** Subsidies of renewable SPPs can be most broadly and cost-effectively implemented with a competitive-bidding scheme to allocate and determine the amount of subsidies. Competitive bidding can also be used to ration and control the award of SPP entitlements.

- **Perpetuating SPP program momentum.** When certain land resources or land-use permits that are necessary for renewable SPP development (for example, hydroelectric or wind) are under government control, there is an interest in preventing speculative hoarding of both the sites and permits or the LOIs to control those sites. A variety of bid security and milestone requirements can minimize such hoarding by extracting an economic rent for holding those entitlements or by requiring progress by certain elapsed times.

- **Embed incentives in the PPA relationship.** Several fundamental choices are available in designing the PPA to address and regulate the buyer–seller relationship between the SPP and the utility. Embedding financial incentives in the tariff and delivery provisions of the PPA can creatively shift aspects of the legal relationship away from legal sanctions and penalties and toward internalized, self-effectuating market incentives.

- **Update avoided cost principles.** A fundamental issue that must be addressed in every system using long-term contracts is whether and how the SPP will be paid for the capacity value of its power. Although some nations have elected to take capacity without paying for it, this diverges from avoided cost principles and creates tension in the long-term utility relationship with SPP developers. Some of these programs have developed creative methods to provide capacity incentives in the PPA tariff structure or to have floating capacity-value payments based on peak-period delivery performance. The PPA tariff structures can be differentiated to provide unique or tiered SPP power sale prices that are based on a variety of SPP variables of value to the system, including the baseload nature of the resource, capacity factor, availability, location on the grid, environmental attributes, and other factors.

- **Implement the proper solicitation mechanism.** Programs can alternate between open SPP power solicitations (similar to the traditional PURPA model) or controlled competitive solicitations as a means of encouraging or controlling SPP proposals. Controlled solicitations for SPP projects can create a hierarchy or tier of desirability for SPP projects by fuel source, prime mover, environmental attributes, location, system requirements or resource–prime mover diversity, ensuring that full capacity requirements are subscribed in the order of highest value or utility to the national supply plan.

- **Evaluate innovative measures.** The future of SPP programs will involve a combination of innovative program design elements to enable private-sector credit support for SPP investments: SPP retail sales, retail wheeling, energy banking, and net metering. Experimentation with these elements has already begun in Asia. These creative elements provide sale and banking alternatives for what is otherwise a perishable commodity sold to a single buyer. This allows independent power to move to its most efficient and highest-value use. However, these options must be carefully structured so that the utility receives fair value for its transmission and banking services and so that its retail power service obligations are not inordinately compromised by these innovative elements. Careful SPP program design and intelligent PPA design are essential to navigating this challenge.

It is clear that there will be a dramatic and rapid increase in the electrification and construction of electric-generating capacity in developing nations. This will occur as predictably as the moon and the tides. The question is, how will the international community and the host countries manage and direct this expansion of electric-generating capacity? The very fact that it is happening rapidly creates opportunity for error. However, the lessons from these five Asian nations provide the information that is necessary to create a successful template for SPP and renewable programs in developing nations. Existing programs and PPAs, in every case, can be improved to better achieve program objectives. A review of existing programs at this point in time, as well as the careful design of new programs, will pay significant benefits to the host countries, multilateral agencies, and the global community.

To succeed, each SPP program must attract international capital to the electric sector. International capital flows predominately from U.S. and European lenders, as well as from multilateral international and regional development agencies. As experience is gained with SPP programs, the program designs and elements that are acceptable to this finite group of lenders can be refined to a predictable array of financial, environmental, and contract requirements. These requirements will become the eye of the financing needle through which successful SPP projects must pass. The following chapters navigate these international financing requirements.

Notes

1. Ferrey, Steven. 2005. *The law of independent power,* 22nd ed. St. Paul, MN: West Publishing. See chap. 9 for a discussion of power bidding systems under Public utilities Regulatory Policies Act (PURPA) in the U.S. model.

2. Ibid., chap. 9.

3. Ibid., chap. 10.

4. Ferrey, Steven. 2003. Nothing But Net: Renewable Energy and the Environment, MidAmerican Legal Fictions and Supremacy Doctrine, and Policy. *Duke Environmental Law Journal.* 19:1.

10 Financing the Transition

Transitions to Restructured Markets

Developing nations' energy sectors are different from those in some developed nations. Publicly owned utilities historically have monopolized the energy sector in most nations and in almost all developing nations. During the 1980s and 1990s, a paradigm shift occurred: The energy sector shifted from a state-owned-monopoly model to a market-based sector model.

- During the 1970s and 1980s, public utilities experienced increased production costs, primarily because of consumer protection and environmental regulations. This cost increase coincided with the development of smaller generating units—typically small gas turbines—that could generate electricity at lower prices than the larger plants traditionally constructed by traditional public utilities.

- In the United States, beginning in 1978, qualifying facilities and independent power producers (IPPs) created independent power projects to connect to the grid; the energy sectors of Chile and the United Kingdom were successfully privatized during the 1980s.[1]

- During the 1990s, the World Bank Group (WBG) decided that the private sector was the solution for energy investments. It made a structural readjustment of its lending policies, reducing the bank's lending support for public-*sector* energy projects from approximately 25% to less than 10%.[2] Other development banks followed suit.[3]

Moreover, publicly owned and operated energy-sector projects often failed to support economic and social development in developing countries.[4] Problems occurred with international subsidy diversion to inappropriate individuals or groups, inadequate utility management, poor project choices, and corruption that inhibited sector development.[5] Furthermore, the growth of electricity consumption in developing countries exceeded the financial resources of state-owned utilities, which often were run at a loss to subsidize consumption, and did not provide adequate revenues to develop electric capacity.[6]

The new market-based sector paradigm involves two components: (1) a change in sector-management practices to favor privatization and (2) a restructuring of the energy market to foster competition and commercial investment.[7] Sector commercialization involves the unbundling of vertically integrated utilities into their individual components, in conjunction with sector reforms focused on creating a competitive market for energy transactions. Successful sector reforms create a market framework by defining legal rules, setting tariffs, and defining contractual terms that favor commercial competition. Ownership privatization is also a necessary component of this transition.[8]

Only by engaging in market-based reforms can developing countries attract sufficient international capital and foreign investment to finance their needed electric-sector development over the next decades. As the previous chapters have highlighted, to successfully adapt, developing countries must commercialize their generating sectors, reform tariffs to reflect avoided cost or similar tariff principles, develop a transmission network, enact a legal framework that allows market unbundling or privatization, and establish an independent regulatory body.[9]

The Magnitude of Required Electric-Sector Investment

Between 1990 and 2000, world foreign direct investment in all sectors of developing countries rose from $962 billion to $1,162 billion (all figures are U.S. dollars).[10] From 1990 to 1999, there was approximately

$187 billion of private investment in energy and electricity projects in developing and transitioning countries.[11]

Despite this overall increase in direct foreign investment, developing countries still face inadequate energy-sector funding because of the rapid rise in demand. Private investment in the power sectors of developing countries has fallen dramatically since 1997, when investment peaked at $50 billion. In 2002, private investment was only $7 billion.[12] Similarly, over the last two decades, the WBG's financial commitments to developing countries' energy sectors have declined significantly. Between 1990 and 1994, WBG energy-sector financing accounted for approximately 15% of the bank's overall commitments. By 2001, however, the annual commitment percentage had fallen to 5% to 8% of overall financing.[13]

The decline in international financing is attributable to several factors:

- A majority of electric-sector investment during the early 1990s went to countries with favorable investment profiles where host governments had embraced market-based sector reform (particularly China, Brazil, IPP programs in East Asia, and the privatization of state-owned utilities in Latin America).[14] However, some countries had difficulty sustaining the market reforms necessary for power-sector commercial viability, making investors more cautious in financing transactions.[15]

- Investors were particularly discouraged by numerous instances in which host governments failed to honor power purchase agreements (PPAs) or carry out promised market reforms. For example, in Indonesia, the government defaulted on sovereign guarantees on two projects, resulting in a loss of $575 million to the project sponsors. Ultimately, despite a ruling in its favor by the United Nations Commission on International Trade Law, the project sponsors recovered nothing from the Indonesian government, but ultimately recovered $290 million under its U.S. political risk insurance policy, which ultimately was funded by U.S. taxpayers.

- Multiple currency crises that occurred during the late 1990s, particularly the 1997 Asian financial crisis, were another significant factor in this downturn.[16] Most IPP capital and running costs for electric power projects, however, were paid in foreign currency and were not devalued. The financial crises led to severe devaluations of local currencies, which had to be converted to foreign currencies to pay for power sales. This undermined the stability of electric-sector investments. As the frequency of the currency crises increased, international capital markets penalized and discouraged investments in developing countries.[17]

- Beginning in 1993, the WBG imposed certain sector reform measures as a condition of electric-sector project financing. The bank's "safeguard policies," essentially an environmental and social review process (see chapter 13), led investors to seek other financing sources.[18]

- Investment flows for electricity upgrades and development were diverted from Asia after the 1997–98 Asian financial crisis. Approximately 50% of total private investment flows were diverted to Latin America and the Caribbean, whereas South Asia and sub-Saharan Africa received only 5% and 3% of total private investment flows, respectively.[19]

Reduced foreign investment forces developing countries to independently fund energy-sector development, limiting domestic savings and diverting funds from other social programs. Moreover, the International Energy Agency (IEA) forecasts that electricity demand in developing countries will increase approximately 4% per year over the next 20 years.[20] Many of the developing nations profiled in previous chapters have much higher growth rates. To keep pace with this demand, the IEA estimates that developing countries need to invest $120 billion per year in the power sector during 2001–2010.[21] This investment, a significant portion of all international energy investment, is required in all aspects of the energy sector: increased generation capacity, expanded transmission and distribution networks, and maintenance of the existing infrastructure. Developing nations will have to attract significant private-sector debt and equity investment to be able to adequately address this need.

Funding Sources

Debt financing

A primary source of funds for international power projects is commercial debt. In a debt transaction, the lender provides funds to the project sponsor in return for an agreement to repay the funds with interest over a fixed term. Power project sponsors typically seek financing terms of 15 to 20 years. As discussed in previous chapters on the individual Asian country programs, in some developing countries, private commercial financing is unavailable for sufficient length or terms.

Foreign commercial banks, local commercial banks, export credit agencies, insurance companies, public debt issues, and multilateral financing institutions are all sources that provide potential loans and debt for project financing. Commercial financing, however, is often available only to established companies with a strong asset base, cash flow, and debt capacity. These requirements create financing barriers for upstart renewable energy projects, limiting sector development. Instead, small power projects are often financed privately with debt secured by the investors' private assets.[22] For projects financed by the International Finance Corporation, a WBG entity, commercial debt provides approximately $16 billion per year, 61% of total lending.[23]

Equity participation

The other major source of project funds for small power providers (SPPs) is equity financing. With equity financing, the source provides the funds to the project in exchange for an equity—or ownership—share in the project. Project sponsors, governments, contractors, specialized infrastructure funds, and multilateral financing institutions are all potential equity sources for SPP project financing.[24] Project sponsors obtain additional capital from venture capitalists and equity investment funds, which contribute to project financing in exchange for partial equity ownership. This type of capital is known as *risk capital* because the purpose of the investment is to provide early-stage seed capital and to obtain a high rate of return if the project is successful. United Nations Environment Programme (UNEP) research indicates that risk capital is a primary form of financing for renewable energy projects.[25]

Grants provide additional capital to power-sector projects. Grant capital is particularly favorable because sponsors need not repay the grant, and it reduces the amount of other conventional financing that must be secured by project sponsors. Host government and public-sector entities are the primary distributors of grants to energy projects, especially in the early stages of project development. When investors seek to acquire a publicly owned utility, they need not engage in an outright purchase from the state. Rather, an innovative means of project financing is a debt/equity swap. An investor engages in a debt/equity swap by purchasing, at a discount, existing foreign country bonds originally issued to fund the state utility. The investor then transfers these bonds to the state in return for equity in the existing utility. This allows the state to retire foreign debt while privatizing its energy sector. Thus debt/equity swaps are advantageous because privatization and improvements in the host country's fiscal status each improve a country's risk profile, as well as the project risk profile, and result in greater investment and lower financing costs, respectively.[26]

Electric-Sector Investment Criteria

Foreign investors are particularly selective when financing projects in developing countries. In 2003, 44 energy-sector firms participated in a WBG survey to identify conditions in developing countries that investors perceived as important when making new investment decisions or determining the performance of current projects.[27] Several revelations emerged from survey responses. The survey demonstrated that investors generally considered the same factors in making initial investment decisions and evaluating the performance of ongoing investment projects in the electric sector. The survey revealed that despite a downturn in overall investment in the energy sectors of developing countries, investors were not uniformly dissatisfied with their investment experience (fig. 10–1).

> The World Bank Group survey[1] indicates that there is a consensus as to what investors want from a host-government in a developing country.
>
> - Investors agree the government must ensure an adequate cash flow to the project by supporting agreed-upon and pre-set tariff levels and supporting the collection of outstanding debts. Investors favor payment discipline programs because they tend to ensure an adequate cash flow to the project. Where a country's customers have a history of nonpayment or the country has a history of lax enforcement, the investment is considered more risky and to reduce this risk, investors in a local utility seek the ability to disconnect customers. This can preserve an adequate cash flow to the project. Alternatively, a government or multilateral agency can provide a guarantee to cover nonpayment and lax enforcement.
> - Countries where the government maintains the stability and enforceability of laws and contracts are preferred by investors. Investors favor developing countries with transparent legal frameworks that specifically define the rights and obligations of the investor, government, and consumers. Countries that have failed to abide by contractual obligations are considered riskier investments.
> - Investors favor administrative efficiency, especially during the bidding and construction process, and minimal governmental interference. Governmental inaction results in lost opportunity costs. Governmental interference may result in lost profits.
> - The availability of credit guarantees is a significant factor in initial investment decisions. Guarantees reduce project risk, enhancing likely returns.
>
> [1] *Id.* at 4-12.

Figure 10–1: 2003 WBG investor survey.

The WBG survey indicated that when respondents invested in developing countries, a substantial percentage sought a return on investment greater than 16% per year and the opportunity to diversify into new markets.[28] When investment experience was evaluated on a regional scale, respondents indicated they had had positive experiences in East Asia, the Pacific, Latin America, and the Caribbean. Investment experiences in South Asia were considered negative.[29]

The survey also yielded several surprising results. The survey concluded that investors did not consider the availability of domestic financing as important as other factors in the investment determination. However, some observers consider local currency funds essential to attracting foreign investment. Moreover, respondents agreed that a country's transition to a competitive market structure is relevant to an investment decision only when the elements for a transparent legal system, independence from regulatory authority, and payment discipline are actually present.

Electric project investors generally favored countries with legal frameworks that define investors' rights and obligations, have and enforce payment systems, or offer a guarantee from the government or a multilateral agency. The same investors also stated that the retail tariff level and collection discipline, adjudication of tariff adjustments, project operational control, and long-term support are essential to overall long-term electric project investment success.[30]

Phases of Financial Risk

Here we analyze the primary types of risks encountered in international power transactions and the mitigation options available to reduce these risks. When evaluating a project, a financier will assess each major source of project risk and determine whether and how that risk can be shifted to other stakeholders or otherwise mitigated. Only when the risks can be mitigated to a level that is within the tolerance of financial-sector participants will project financing move forward. Typical mitigation measures involve legal and contractual provisions that shift risk (see chapter 11). These measures are only as successful as the ability to legally enforce contractual provisions and guarantees.

Power projects that are bolted into the ground in a foreign land are especially exposed to long-term risk because the capital-recovery horizon is long term and linked to the periodic earning of revenue from the sale of power output. The facility itself, once constructed, cannot be relocated. International power projects are also considered risky because the margin between revenues and costs may be thin. In fact, the average retail energy prices in developing countries (3.8 cents/kWh) may actually be lower than the average cost of generating the electricity (4–10 cents/kWh) if subsidies, tariffs, and other sources of revenues are not considered.

From the lender's perspective, there are three temporal project risk phases: (1) design, engineering, and construction; (2) start-up; and (3) operations. The risks during each phase are distinct and require separate mitigation measures. As chapter 11 discusses, during all stages, the provisions of the PPA, as well as private or multilateral insurance policies, are critical to shift and mitigate the risk of loss.[31]

Risks arising during design and construction

Design, engineering, and construction risks are of special concern to investors because they occur after the loan is made but before the project begins to generate revenues. Significant design, engineering, and construction risks include cost overruns, time overruns, and site availability. These risks may be mitigated by contractual and legal provisions, such as escalation agreements to protect against overruns, performance obligations to warranty production, and land-use agreements.[32] Moreover, insurance instruments are available to cover instances of physical loss or damage during the design and construction period, as well as delayed operational start-up because of constructions delays. Such insurance also may be used to cover contractor liability.[33]

Risks arising during start-up

Risks during the start-up phase of a power project include technology nonperformance, inputs pricing, and the unavailability of skilled labor. The risk of technology nonperformance, which is especially relevant for renewable energy projects, may be mitigated by securing performance bonds or guarantees for technology performance from the manufacturer. To protect project profitability from an increase in inputs pricing, investors may execute supply contracts with pass-through provisions. This allows projects to pass certain cost escalations on to consumers. Finally, to ensure the availability of skilled labor, investors may require that the project include labor training programs.[34]

Operational risks

Once the facility is operational, lenders still face numerous project risks, including risks of project performance, payment and collection risks, and noncommercial risks. Project performance risks may be mitigated by contractual risk-allocation agreements. Payment and collection risks may be mitigated through the use of escrow accounts, standby letters of credit, and government guarantees.[35] The risks discussed in the next section—sovereign risk, expropriation risk, and foreign exchange risk—typically arise during project operation.

Types of International Project Risk

Multiple types of risk are inherent in the international financing of power-sector projects in developing countries. These include the risk that the host government will fail to fulfill certain contractual obligations (sovereign risk), the risk that the host government will deprive the project investors of debt and equity rights in the project (expropriation risk), and the risk of currency devaluation and availability (foreign exchange risk).

Sovereign risk

Sovereign risk is the risk that the host government or national utility will not fulfill a promise to purchase power from the project. Examples of sovereign contractual risk include the host government's failure to maintain an agreed-upon regulatory framework (including tariffs), failure to collect payment for power outputs, or failure to compensate for delays in government action.[36] Sovereign risk also arises when the government owns or controls access to fuel reserves needed for power generation. Similarly, sovereign risks include instances in which the host government fails to adhere to energy-pricing tariff determinations, changes in local regulations that are applicable to the project, and changes in foreign ownership rules. When a regulatory or interpretive change of law occurs, such changes are often called *policy risk.*

When evaluating sovereign risk, the host government controls the legal system, the project sponsor's primary means of recourse. Examples of host government failures are discussed in figure 10–2. These examples serve as a reminder that risk cannot be entirely eliminated in an international power transaction.

- *Hubco Project, Kurachi, Pakistan:* In 1995, HUBCO finalized the financial arrangements for an oil-fired project in Pakistan. Prior to operation, HUBCO and the state-owned Water and Power Development Authority (WAPDA) had executed a power purchase agreement. WAPDA subsequently disputed the contractually agreed-upon power purchase price and refused to pay, resulting in $125 million annual revenue shortfall for the project. Pakistan refused to honor and uphold the sovereign risk guarantees previously provided by the government to HUBCO. This forced HUBCO to renegotiate with the utility, WAPDA. The renegotiated agreement resulted in a tariff price of one cent/Kwh less than the former purchase price set by the contract, a significant reduction.
- *1998 Indonesia Financial Crisis:* During the 1998 financial crisis in India, PLN, the nationall utility, cancelled contracts with twelve power projects. PLN then stipulated new, and lower, power purchase prices for each project. Local Indonesian courts provided no recourse for the project owners.

Figure 10–2: Examples of host government failures.

Expropriation risk

Expropriation risk is the risk that political actions of the host government will deprive investors of equity and debt rights in the project. This could be a partial or full expropriation or nationalization of the capital asset or the power output of the project. Alternatively, the host government could engage in constructive expropriation (the use of abusive tax measures in combination with injunctions), abuse regulatory powers to cause harm to project sponsors, or target trade restrictions specifically to the project.[37] When negotiating contracts with host governments, investors can minimize expropriation risk by ensuring that the contract includes provisions that require disputes to be judged by international law (see chapter 12), includes a fair and equitable treatment clause, provides for unrestricted transfer of funds to and from the host country, and requires dispute resolution outside the host government's judicial system.[38] However, in some relevant instances, host governments have refused to observe rulings by international tribunals. Moreover, it is notable that expropriation risk does not include actions that the government takes as a purchaser of power, supplier of fuel or other key project inputs, creditor, or shareholder of the project.

Foreign exchange risk

Foreign exchange risk is a significant factor in private investment decisions.[39] Foreign exchange risk arises when the host country lacks developed capital markets and cannot provide local financing for a sufficient term. To secure financing, investors are forced to seek international financing and investment. As a result, project revenues are denominated and paid in local currency, whereas project expenses, including loan repayments, are paid in a foreign currency, typically the U.S. dollar, Japanese yen, or the euro.

Exchange rates between currencies do not remain fixed over a project's life. Typically, inflating economies in developing countries devalue their currencies over time compared to the reference foreign currency of the dollar, yen, or euro. This devalues the exchange and value of local currency relative to the stable reference currency if the PPA tariff is denominated in local currency.

Power projects are uniquely subject to foreign exchange risk because power output is not tradable in international markets. Furthermore, it is not portable when the transmission system for accessing such trades is controlled by the local utility, and, once assets are installed, they cannot be redeployed. This compels project operation to continue, even in risky situations. Moreover, power projects are acutely exposed to foreign exchange risk because power plants are typically depreciated over 20 to 30 years, creating increased exposure to risk because of the length of the investment life.[40]

There are three components of foreign exchange risk: exchange rate risk, convertibility risk, and transfer risk. The primary component of foreign exchange risk is exchange rate risk, which typically arises when project revenues are denominated in local currency, when some project expenses are denominated in a more stable foreign currency, and when there is depreciation of the local currency. When currency depreciation occurs, there is a reduction in project revenues when the revenues are converted from the local currency, often leaving insufficient funds to cover project expenses.[41] Another way of understanding exchange rate risk is to consider it the variability in the value of a project (or a portion of the project) resulting from changes in the exchange rate. Both the project value and financing are subject to exchange rate risk. Project exchange rate risk occurs when the value of the project's inputs and outputs are dependent on the exchange rate.

Convertibility risk, the second component of foreign exchange risk, is the risk that the project investor will be prohibited from converting local currency revenues into foreign currency because of local government policy decisions regarding the exodus of revenues generated by activities in the country. These restrictions are often enacted for macroeconomic reasons rather than power project economics. Similarly, transfer risk, a final component of foreign exchange risk, is the possibility that the owner will be prevented from electronically or manually transferring revenues out of the country, even if the currency can be converted. Transfer risk and convertibility risk are distinct.

Shifting and Bearing Risk in the Power Sector

In a functional market economy, the party that is best able to bear the risk is the party to which risk should be allocated.[42] In energy-sector development, there are three primary parties to which risk can be allocated: investors and lenders, consumers, and the host government. Thus, a key part of project financing negotiations is contractually determining which party will bear which risk for what period of time. The ultimate allocation of risk is embodied in the power purchase agreement and in any ancillary, consistent loan agreements (see chapter 11).

Allocation of foreign exchange risk

No one party has direct control over foreign exchange risk. Moreover, there is no consensus as to which stakeholder should bear the brunt of the foreign exchange risk.[43] One distribution is to allocate exchange rate risk based on the party's ability to influence the exchange rate, ability to influence the value of the project (by changing demand or expenses), and ability to hedge or diversify the risk. Generally, investors have a greater ability to diversify risk, customers have a greater ability to influence project costs, and governments have a strong influence over exchange rate fluctuations.[44]

Some argue that the host government is best situated to bear foreign exchange risk because it controls monetary and fiscal policy.[45] Allocating substantial risks to the host government, however, reduces the country's risk profile, potentially reducing its likelihood of attracting future foreign investments. Allocating exchange rate risk to governments is not recommended because, although they control financial policy, they typically do not respond well to financial incentives.[46]

Others argue that project investors are best situated to bear foreign exchange risk because they can use currency-hedging devices and vehicles to mitigate the risk. Currency hedging is a process that firms use to offset foreign exchange exposure in the event of a currency devaluation. One hedges currency risk by entering into futures transactions that realize profits when the local currency depreciates against stable currencies. Futures contracts, agreements to purchase a certain sum in local currency at a set exchange rate, are the most common form of hedging. When currency-hedging opportunities exist, investors are well positioned to bear foreign exchange risk. Often, however, derivatives markets are nonexistent in developing countries, making currency hedging unavailable. This is true in some of the countries profiled in earlier chapters.

Finally, it has also been suggested that consumers and citizens bear the foreign exchange risk. Indeed, consumers are the most diversified class, thereby minimizing risk as it is spread across many individuals. Moreover, allocating this risk to consumers forces them to pay the full cost of energy supply, hypothetically resulting in a pure demand curve, where consumers adjust their energy consumption accordingly.[47] Such reasoning, however, may not transfer into the real world. Moreover, this argument ignores consumer welfare. As discussed in chapter 1, many low-income households cannot afford connections to the transmission network. Forcing them to also bear the costs of substantial risk further discourages access opportunities.

Thus, there are several options for the allocation of these risks. Project foreign exchange risk can be allocated differentially between consumers and investors, with the greater percentage born by the party that is best able to diversify the risk and influence the value of the project. Of the two parties, investors may need to absorb a majority of the financing related to exchange rate risk because they determine the type of financing and have a greater ability to diversity risk.[48]

Allocation of sovereign risk

Sovereign risk is generally allocated contractually between project sponsors, financiers, and the host government. In executing a contract with the host government, project sponsors attempt to extract as many obligations as possible so as to have enforceable commitments. Similarly, when financiers issue guarantees, which pay on the occurrence of a certain event, they also will require the host government to sign a counterguarantee or indemnity agreement. Thus, as a contractually responsible party, the host government often has an incentive to minimize sovereign risk. These guarantees are only as strong as the PPA, the guarantee itself, and the ability of courts to enforce these guarantees. (The PPA is discussed in chapter 11, and the rules for enforcing legal provisions are examined in chapter 12.)

Risk Mitigation

Credit rating agencies evaluate the degree to which a specific project and a specific country is subject to various types of risks. When the project profile or country profile is too risky, financing is unlikely or will be available only at high interest rates with multiple guarantees. A project sponsor or host government can take measures to improve its risk profile.

To diminish sovereign risk, investors may obtain a guarantee from the host government for the utility revenue stream or use escrow accounts when utility revenues go directly to escrow and then are transferred to investors. Alternatively, investors may obtain a guarantee from a multilateral bank, which guarantees payment from the bank on the occurrence of specified events. Use of any of these mechanisms will improve project creditworthiness. Risk guarantees are discussed in more detail later in this chapter.

Numerous strategies can be employed to mitigate foreign exchange risk. The methodology for mitigating foreign exchange risk is dependent on country-specific conditions and the type of financing required. This section discusses several of the most common strategies for improving the creditworthiness of international power project transactions.

Developing local capital markets

By developing local currency financing markets, project investors can obtain full or partial local currency project financing of the requisite maturity. This allows investors to obtain financing in the same currency in which revenues are paid, eliminating the risk to the project from currency fluctuation. Instead, the risk of currency devaluation is allocated to local financers.

Ultimately, the success of financing a project in local currency is contingent on the development of long-term, fixed-rate energy project financing in developing countries. Currently, there are numerous constraints on local capital market development, including the absence of long-term government bonds to provide a basis for financing long-term debt and the lack of financing experience. Moreover, country risk-exposure profiles may prevent international support of local currency markets.[49]

Local currency fund schemes

When local capital markets are not well developed, an alternative is to develop a local currency fund scheme to provide long-term local currency financing and to diversify project risks. A fund scheme functions by obtaining initial capitalization from project investors, institutional sponsors, and the host government. This capitalization is then used to create a reserve fund from which local currency bonds are issued. The bond issuance spreads the project risk over a new pool of investors, and revenues from the bond issuance are loaned to the project.

Partial credit guarantees: Local currency credit enhancements

Partial credit guarantees (PCGs) are often used when local capital markets cannot provide financing of requisite length. Generally, PCGs are available to cover all of the credit risks inherent in a specified portion of the financing or to cover events with later maturities. By covering these risks, PCGs allow investors to obtain financing with maturities beyond those commercially available, with typical maturities of 15 to 20 years.[50] Partial credit guarantees also may be used to ensure the continuity of financing. The use of PCGs is expected to increase to ameliorate the decline in private investment in developing countries.[51]

Partial risk guarantees and political risk insurance

When local financiers are willing to finance a power project but are deterred by the long-term risk, a partial risk guarantee or political risk insurance can be employed to protect financiers against political and regulatory risks. A partial risk guarantee provides an insurance backstop on the occurrence of a specific event. Similarly, political risk insurance provides insurance on the occurrence of a specific political action by the host government. That trigger event might be a change of law or political actions that disrupt project cash flow. Although insurance instruments can alter some risk, successful macroeconomic performance of the host country is the surest way to improve risk factors and attract foreign capital.

Comprehensive guarantees

An international bank or institution can provide a comprehensive guarantee of local power project financing. Underlying a comprehensive guarantee is a local loan that is issued to the power project and guaranteed by an international financing institution. The guarantee assures the local bank that the financing will be repaid by either the project or the guarantor.[52]

Risk-mitigation tariff formulas

An alternative technique for mitigating risk, particularly foreign exchange risk, is to include an indexation formula in the PPA tariff agreement that adjusts prices to account for variability in certain project- or financing-related risks. This agreement is contractually developed between the project investors and the host government. Indexation is favorable to project investors, primarily because it allows project revenues to be adjusted to reflect an indexed exchange rate. Similarly, it allows investors to pass certain foreign currency costs on to consumers. Indexation formulas, which index price directly to the exchange rate or to inflation, however, may be unfavorable to local government fiscal policy. Typically, the formula adopts a hybrid approach, with only some of the cost (those capital or fuel costs that must be paid in foreign currency) indexed and some not.[53]

Government Role and Reform for Small Power Producer Programs

There is a consensus that public financing of energy-sector projects has failed to support economic and social development in developing countries.[54] This is particularly true with respect to impoverished communities located in off-grid regions. In many developing nations, there is a need for private capital to finance electric infrastructure development. Numerous countries have attempted to attract and implement renewable small power project development. These projects seek to attract private-sector capital to augment power development in the nation.

Lessons for a successful small power producer model

Some market design flaws that occurred in the California deregulated market serve as important lessons for developing countries that seek to restructure or privatize their electric sectors. The problems that California experienced with restructuring its electric sector, deregulating wholesale markets, and facilitating independent power can be categorized as follows:[55]

Market Design Flaws
- Deregulation of the wholesale market while retaining a regulated price in the retail market
- Lack of adequate economic or siting incentives for capacity expansion and generator supply construction
- Demand inelasticity because of a retail price freeze and a perception of electric power as a necessity by consumers
- Inconsistent price caps on the prices that wholesale generators could charge for wholesale power and arbitrage by generators in an undersupplied market
- Lack of independent oversight and governance by independent regulatory or stakeholder groups
- Lack of retail competition (in California, less than 2% of all retail customers migrated to alternative suppliers)

Exogenous Factors
- Constraints on expanding supply because of environmental controls, difficulty obtaining siting permits, and market uncertainty
- Uncertainty and a decrease in importation of power from neighboring states
- Unreliable older generating and transmission facilities
- Unexpected increase in air conditioning use and resulting increases in retail demand
- Lack of reliability and resulting blackouts of retail customers
- Unexpected increase in the cost of fossil fuel for certain fossil-generating units
- Inability to operate certain fossil fuel-fired generating facilities that had exceeded their environmental emission limitations for the year or the quarter

The host government must be an active participant in all phases of the energy-sector financing transaction. Initially, the host government must be active in identifying sector needs and determining what types of investment it desires to stimulate the sector. If foreign investment is desired, the government must select where to direct the investment and promote and develop policies to encourage this funding. As discussed previously, this typically involves some degree of market privatization and commercialization. Once these reforms have been implemented, the host government must strive to ensure market and regulatory stability. Often, the host government completes each of these assessments with the help of a private partner, such as the WBG, United Nations Development Programme (UNDP), or UNEP.

Some of the SPP programs in developing countries have been initiated in response to funding opportunities from the World Bank or other multilateral funding authorities. These targeted efforts have provided technical assistance to the host countries for the development of a structure for the SPP program, analysis of tariff issues, assessment of local lending opportunities, and development of a standardized PPA and standardized tariff structure for SPP implementation. If these issues are not developed in a standardized, impartial, and neutral manner, they can be extremely contentious among various stakeholders, and the program can fail.

Public–private partnerships also may be implemented. World Bank Group studies, however, indicate that it is difficult to establish such agreements, and their effectiveness tends to diminish after the initial management privatization.[56] An alternative to management partnerships is outright power service concession and asset transfer. This involves direct sales of state-owned utility generation facilities or transmission networks to private utilities, which then assume the obligation to service a certain geographic area.

Reform of the Existing Legal Framework

To promote sector commercialization and privatization, the host government must reform its existing legal framework. Such reforms require the development of a competitive market system, in some cases a competitive-bidding framework, a legal framework to support contractual arrangements and the PPA, and the creation of an independent regulator and regulatory oversight.[57] A competitive market is necessary to attract international developers. As part of that competitive market, there must be a fixed tariff for long-term power sale that is sufficient to attract new market entrants. The prices for power generation projects will be bid down with competition for the right to earn the entitlement to build and own a project. Competitive bidding, as Thailand employs for its renewable energy subsidies, is a means of rationing the selection of specific projects or cost-effectively spreading subsidies over the broadest possible base of additional generation assets.

Equally important is the legal structure of the developing country. This includes the willingness to enter into neutral and balanced PPAs, a tradition of prompt enforcement of contracts in neutral forums, and fair decision rules for contract adjudication. (The PPA and the application of PPA decision rules are discussed in chapters 11 and 12.)

Many developing nations have a single buyer for independently produced power: the state-owned or state-regulated electric power company. This single-buyer model often postpones an essential element of any privatization and reform, the setting of retail prices at levels that will cover wholesale system costs. If subsidized consumers are insulated from the true price of electric power, even temporarily, they will end up paying for this inefficiency because less money will be available for other sectors of their economies, such as hospitals, roads, and schools, which will not get the subsidies that are being diverted to the electric power sector.

Creation of an independent government regulator

Government must remain involved in its commercialized energy market. Not only does the government have direct jurisdictional control over the utility monopoly in many developing nations, but also the government sector is the primary agent for guaranteeing a fair and balanced privatized system in nations with immature judicial sectors. Potential government roles include regulation to protect the public interest (against monopolies and environmental degradation), creation and administration of incentives for private facilities to protect the environment and to increase access to low-income populations, reduction and mitigation of political risk, and reduction of barriers to market energy for new sources. The creation of an independent regulator is critical for several functions. Creation of an independent regulator shifts the following elements of the electric sector:

- It can approve PPAs that are balanced and fair, limiting the monopsony power of the utility.

- It can set the tariff so that the power-buying utility does not do so unilaterally.

- It can oversee SPP program design and implementation.

- It can entertain, mediate, and resolve disputes between SPPs and utilities.

- It can develop policy initiatives to make SPP programs successful.

- It can determine and enforce the rules of evolving competitive markets.

It is important for the regulator to have the independence to take positions that are independent from, not controlled by, the government or the utility. In many developing nations, the centralized role of the government in developing energy policy and the very primary role of the government-controlled utility monopoly requires that competitive program design be overseen and administered by an independent agent.

International Credit Agencies and Their Financing Functions

World Bank Group

The World Bank Group is a specialized agency of the United Nations and a conglomeration of multiple entities, including the International Bank for Reconstruction and Development (IBRD), International Development Agency (IDA), International Finance Corporation (IFC), and Multilateral Investment Guarantee Agency (MIGA).

The amount of annual lending from each of these agencies decreases in the order in which they are listed. Each bank specializes in the financing of different aspects of projects, based primarily on the type of financing provided. The WBG is funded primarily through borrowings in the international capital markets and is one of the largest sources of financing for energy-sector projects in developing countries.[58] The cumulative lending of these four World Bank entities is about $0.5 trillion, with current annual lending of about $23 billion.[59]

The IFC loans mainly to private-sector companies operating in developing countries, whereas the IBRD and the IDA make loans to country governments. During the last decade, the share of lending to the energy sector has fallen from approximately 25% to less than 10% of all IBRD and IDA lending.[60] With this precipitous decline, currently only about 5% of total loans go to the electric power sector; of this, approximately 55% go to fossil-fueled power generation facilities, 20% to transmission and distribution, 10% to oil and gas activities, and about 15% to sustainable energy projects.[61] Most of the renewable projects that are funded are small- and medium-scaled hydroelectric plants and some photovoltaic facilities and biomass programs, with a small portion going to wind and geothermal renewable energy.

The WBG entities can finance projects jointly.[62] Moreover, all of the entities follow a similar funding process. Phase I involves project identification. Phase II involves the development of project agreements and financial arrangements, including the negotiation of bank loans and guarantees. Phase III involves approval of the project on a regional and bankwide scale, followed by approval of the WBG board. During this stage, the projects undergo environmental review, which can be a significant hurdle. (The environmental clearance procedures and requirements of the key international banks are analyzed and contrasted in chapter 13.) In Phase IV, all agreements are executed and monies distributed to the recipients.[63]

Any WBG-supported investment must satisfy at least one criteria of its business strategy framework, which includes direct poverty reduction,[64] country fiscal stabilization,[65] governance and private-sector development,[66] and environmental sustainability.[67] Similar goals are reflected in the WBG's midterm objectives, which are structured to achieve these policies.[68] The rationale for WBG's midterm policy objectives is that increasing access to modern energy services will lead to poverty reduction in low- and middle-income countries.[69] Moreover, expanding energy sectors in the short term provides greater energy availability for economic activity and employment and allows redirection of government funds to poverty-alleviation programs and other social needs.[70]

To achieve these objectives by 2010, the WBG has set a series of quantifiable goals for its investment in developing countries. These goals include increasing the share of households with modern energy access from 65% to 75%, increasing the number of cities with acceptable air quality from 15% to 30%, decreasing the average CO_2 emission intensity of energy consumption from 2.90 t/toe to 2.75 t/toe, and increasing average energy consumption per unit of gross domestic product. Moreover, the WBG seeks to increase consumer supplier choice to 40%, lessen public power-sector investing burdens, and increase the amount of private financing in the energy sector from 25% to 40%.[71]

International Bank for Reconstruction and Development

The IBRD was created in 1945 and currently has 184 member countries.[72] It can provide direct lending, technical assistance, and guarantees. When the IBRD engages in direct lending, the lending terms are set at market rates, with a maturity period of up to 20 years. Generally, IBRD financing is available to public- and private-sector projects.

Loans. One advantage for developing countries is that IBRD instruments often have a three- to five-year grace period before loan repayment begins.[73] If the loan is denominated in local currency, it can be paid back over time in an often-deflating local currency, which lowers the effective interest rate even more compared to borrowing international capital from countries loaning in dollars or euros. Financing may be extended in instances when a public-sector entity is undergoing financial restructuring.

As of June 2003, the IBRD has provided approximately $383 billion in cumulative lending and has loaned approximately $149 billion to countries around the world.[74] In 2003, the IBRD provided a total of $11.2 billion in financing to operations in 37 different countries.[75] Of this, IBRD spent more than $1 billion on energy infrastructure and service development. Moreover, the bank spent $2.5 million to fund legal and policy reforms.[76] The investments of the IBRD and IDA in the energy and mining sectors of South Asia and East Asia are shown in table 10-1.

Table 10-1: World Bank Lending to Energy & Mining Sectors in East Asia/Pacific and South Asia from 1994 to 2003 (in millions of dollars)[1]

Location	1994–97 (annual average)	1998–99 (annual average)	2000	2001	2002	2003
East Asia	1,623.6	517	640.5	142.2	314.5	254.3
South Asia	360.6	545.9	277.8	746.2	504.8	150.6

[1] Adapted from 2003 World Bank Group Annual Report

Guarantees. Because of sector privatization and commercialization, the IBRD has expanded its guarantee program to cover risks that private financiers are unable to mitigate or unwilling to accept. The IBRD offers two types of guarantees: partial risk guarantees and partial credit guarantees. By lowering financing costs and interest rates, investors can attract additional financing and extend the maturity of the loan.[77] Guarantees also reduce the host government's risk exposure.

The IBRD's partial risk guarantee (PRG) covers specified project risks, typically either force majeure events or nonperformance of contractual obligations by the host country (sovereign contractual risk).[78] These include maintenance of the regulatory framework, power project inputs (fuel) delivery, and payment for power. A PRG also may cover foreign exchange risks, with the PRG triggered by currency devaluation. Partial risk guarantees are often used in limited-recourse financing projects, such as build-operate-transfer and build-own-operate power projects.[79]

The IBRD's partial credit guarantee covers all credit risks inherent in a specified portion of the financing. The purpose of a partial credit guarantee is to extend maturities beyond those commercially available or to cover events with later maturities, with typical maturities of 15 to 20 years.[80] This is essential in many developing countries, where private lending institutions have no history of making loans to the energy sector, make only short-term loans, or require the lender to hold a large deposit in return for extending the loan. Partial credit guarantees also may be used to ensure the continuity of financing, that is, for rolling over short-term financing. Investors typically use partial credit guarantees when local lending institutions are involved.[81]

The IBRD guarantees cover both principal and interest, but they protect only debt instruments. Guarantees from MIGA, however, may cover both debt and equity.[82] Guarantee terms and conditions are formalized in a guarantee agreement between the bank and the lender. Guarantee issuance is conditioned on the host country providing a counterguarantee. Under this counterguarantee, the host country must execute an indemnity agreement with the IBRD, making it liable to the IBRD if it fails to fulfill the obligations set out in the guarantee. Guarantees may be repaid in an initial up-front transaction or on an installment basis.

International Development Association

The IDA was created by the World Bank in 1960 and currently has 164 member countries.[83] It is supported by government contributions and provides interest-free credits to countries with the lowest levels of per capita income.[84] In 2002, the IDA provided $8.1 billion in financing for

133 projects in low-income countries.[85] In 2003, the IDA made cumulative commitments of more than $142 billion and provided $7.3 billion to operations in 42 low-income countries.[86]

Multilateral Investment Guarantee Agency

The purpose of MIGA is to encourage foreign investment in low-income countries. It began operations in 1990 as an independent branch of the WBG and now has 162 members.[87] Its instruments provide equity financing, usually in the form of political risk insurance, to insure against noncommercial risk in developing countries. These instruments can provide up to 20-year coverage for certain risks, including currency transfer restrictions, war and civil disturbance, expropriation, and breach of contract by the host government.[88] The agency is willing to cover a variety of investments, including shareholder loans, loan guarantees by equity holders, and commercial bank loans.

One specific advantage of MIGA is that it can coinsure or reinsure other insurance. Also, a MIGA guarantee may cover up to 350% of the total unimpaired subscribed capital and reserves of the project. However, MIGA guarantees have limited project exposure ($50 million) and country exposure ($225 million). Currently, MIGA has committed $12.4 billion in cumulative guarantees. In 2003, MIGA provided $1.4 billion in new guarantees.[89]

International Financing Corporation

The IFC was founded in 1956 with the purpose of financing private-sector projects and providing advisory services. Currently, the IFC comprises 175 member countries.[90] The IFC is legally and financially independent of the rest of the World Bank, although it coordinates its activities with other bank institutions. About 30% of its budget goes to the energy sector, with oil and gas projects commanding a significant portion.[91]

IFC financing supports private-sector projects in developing countries through debt and equity financing, its B-loan program, underwriting, and technical assistance. Private parties use IFC financing to gain access to new sources of funding. Means of project financing include long-term loans, equity investments, subordinated loans, guarantees, standby financing, and risk management.[92]

Loans and instruments from the IFC may provide up to 25% of the total project investment. If the IFC is not the project's largest shareholder, the maximum limit increases to 35% of total equity.[93] Since 2000, the IFC has also been a vocal proponent of market reform and privatization. An IFC strategy is to focus on privatizing distribution and creating regional electricity initiatives.[94] Moreover, the IFC offers a local currency partial credit guarantee, which may be used to reduce foreign exchange risk exposure. The IFC also developed a new partial credit guarantee, which is called on at the occurrence of certain events (typically a significant currency devaluation).[95] The IFC has committed more than $23.4 billion in loans. In 2003, the IFC committed $3.9 billion to projects in 64 countries.[96]

The IFC's guarantee program covers debt instruments (loans, bond issues, and commercial paper) in cases of nonperformance of an obligation by the government for commercial or political reasons. The IFC also offers local currency partial credit guarantees and is currently developing a partial credit guarantee to cover certain debts on the occurrence of specific circumstances.[97] All instruments cover the principle and interest of the outstanding debt.

There are several comparative disadvantages to IFC financing. First, the IFC does not accept counterguarantees by the host government. Thus, financial assistance is provided at the market rate of interest. Second, the IFC requires that the project be privately owned or operated. Finally, the project must meet economic viability and environmental criteria, which are discussed in detail in chapter 13.

Prototype Carbon Fund

In 1999, an international agreement created the Prototype Carbon Fund (PCF), a partnership of public and private stakeholders that share a vision of greenhouse gas reduction. Participating governments include Finland, Canada, the Netherlands, Norway, and Sweden. Private participants include international financing institutions and international energy corporations.

The purpose of the PCF is to help finance and monitor programs designed to reduce greenhouse gas emissions and generate emissions reduction credits, which can be registered pursuant to Article 12 of the Kyoto Protocol. In addition, the fund provides technical assistance to countries that are developing emissions trading programs pursuant to the Kyoto Protocol, trebling its amount of technical training assistance in carbon financing. The PCF makes a market in carbon credits that are trading at about $3 per ton of CO_2 equivalents. Most PCF funds have gone to Latin America, with only 7% to Central Asia, 2% to East Asia, 7% to South Asia, and 18% to Africa. It has funded wind, waste management, bagasse, biomass, energy efficiency, geothermal, small hydroelectric, and photovoltaic technologies.[98] The PCF has approximately $247 million in active pipeline projects.[99] The fund will be extinguished once private-sector markets develop.

For the PCF, one means of reducing carbon emissions is to pay for verified reductions in carbon emissions. Payments are made specifically to projects that have executed emissions reductions contracts with the PCF. As with all international financing transactions, the PCF engages in risk management to ensure actual reductions and a constant payment stream.

The PCF has been active in the development of carbon financing in Asia. In 2003, the PCF completed the identification phase of its carbon financing program for East Asia. The search identified carbon reduction financing opportunities valued at more than $160 million. These projects undergo due diligence, risk management, and contractual negotiations. Moreover, in 2003, the PCF reached carbon financing agreements with projects in Vietnam, Indonesia, China, and the Philippines.[100] Presently, approximately 46% of active PCF pipeline projects are located in East, South, and Central Asia.[101]

Global Environment Facility. The Global Environment Facility (GEF) was established in 1991 by international treaty as a financial mechanism on biodiversity, climate change, and persistent organic pollutants. The purpose of the GEF is to provide financing for non-carbon-based technologies and to create market development incentives favoring renewables.

The GEF is governed by a council of 32 countries, with the largest donors and China having permanent seats. Governing authority is based on a combination of United Nations and World Bank principles, with contested votes requiring 60% of countries, both by number and dollar contribution.[102] The GEF programs are implemented by the UNDP, UNEP, IBRD, and IFC, as well as some of the bilateral donor organizations and regional development banks.

Resources are obtained as pledges from donor countries. The GEF uses these resources to finance the incremental costs of global environmental measures in developing countries. Once GEF approves a specific project, funding is provided by one of three implementing agencies: the World Bank, the UNDP, or the UNEP. Project criteria are guided by the Conventions on Biological Diversity and Climate Change.[103]

Over the last 10 years, the GEF has provided more than $3 billion in grants for approximately 600 projects. Projects have included direct financing of wind power, support of energy-efficiency programs, financing and assistance to rural solar photovoltaic programs, and funding to remedy legal and regulatory barriers need to promote market efficiency and private investment.[104] It administers $2 billion, with a budget increase of $2.9 billion for 2003–06. Climate change activities consume about 36% of the GEF budget.[105]

Energy Sector Management Assistance Programme. The World Bank is dedicated to reducing the carbon intensity of the energy projects that it funds. The Energy Sector Management Assistance Programme (ESMAP) is one of the World Bank's four trust-funded energy programs, with an overall budget of $25–$30 billion. ESMAP funds small-scale projects and deals with institutional capacity building, reviews programs in countries, and conducts energy-environment sector studies. It has been targeting sector reform and rural energy access. Many of ESMAP's renewable energy projects were shifted to GEF. The other trust-funded energy programs are the Asia Alternative Energy Programs, which is concerned with renewable energy and energy efficiency in Asia; the Africa Rural and Renewable Energy Initiative, which is aimed at dealing with rural energy access and renewable energy market development in sub-Saharan Africa; the Regional Program on the Traditional Energy Sector, which assists sub-Saharan Africa in planning and development of the traditional energy sector.[106]

International Centre for Settlement of Investment Disputes. The International Centre for Settlement of Investment Disputes (ICSID), created in 1966, provides a forum for arbitration of disputes between governments and investors. The purpose of the ICSID is to ensure mutual confidence between host governments and international investors. Since 1966, ICSID has resolved 129 cases, including 26 in 2003.[107]

Other International Credit Agencies and Sources

Bilateral donor organizations contribute significant amounts to renewable energy financing in developing countries. Major donors in absolute terms are Canada, Denmark, France, Germany, Italy, Japan, the Netherlands, Sweden, the United Kingdom, and the United States; European countries contribute more than 50% collectively.[108] The German Financial Aid Agency, Credit Kreditanstalt fur Wiederaufeau, and the bilateral development agencies in Denmark, the Netherlands, and Sweden have contributed significantly to renewable energy projects in developing countries.[109]

The U.S. Export-Import Bank, Asian Development Bank, and the Export-Import Bank of Japan all provide credit support. Similarly, the Overseas Private Investment Corporation and the Asian Development Bank each provide certain types of loan guarantees. The most prominent of these agencies and development banks are discussed here.

U.S. Export-Import Bank. Export credit agencies are publicly funded institutions that promote exports from their countries by directly offering credits or by guaranteeing risky credit lines. They use public taxpayer money to provide exporters and their banks with insurance, guarantees, and different types of risk-mitigation measures. Export credit agencies have grown to about $100 billion of credit extension, with about 50% of that targeted to large infrastructure projects in developing countries. The credit support has not been particularly geared to renewable energy project credit guarantees.

The U.S. Export–Import Bank's finance division is a major player in international power transactions. U.S. Export-Import Bank funding supports capital equipment sourced from U.S. suppliers. Therefore, its funding exclusively supports U.S. power equipment sales or U.S. project sponsors. This is a taxpayer-subsidized program that assists U.S. businesses, where other country export-import banks support their nations' power equipment sales.

U.S. Export-Import Bank funding is particularly important because it is available to creditworthy projects even if the host country is already overleveraged or not creditworthy itself. Other additional advantages of the U.S. Export-Import Bank are its ability to cover political risk during construction, to cover financing directly, and to lend for longer terms than commercial lenders. Although the U.S. Export-Import Bank has no caps or country limitations, it does require that equity owners retain ownership for a period of years, that equity participation must remain at risk, and that investors assume first-loss risk.

Overseas Private Investment Corporation. The Overseas Private Investment Corporation (OPIC) is a governmental agency established by the United States in 1971 to facilitate the investment of U.S. private capital in the economic and social development of developing nations. The purpose of OPIC is to provide insurance and project financing to U.S. investors for energy-sector projects in developing nations.[110] The OPIC provides up to 50% of the financing for new projects and 75% for project expansion up to $200 million. The OPIC will insure a project for up to 90% of its investment value, so long as the value does not exceed $200 million. Rates are determined on a case-by-case basis. Instruments of OPIC also can provide coverage for expropriation risk. All OPIC-financed and -insured projects must meet WBG environmental guidelines (see chapter 13).

Asian Development Bank. The Asian Development Bank (ADB) was founded in 1966 and now has 63 members. Its purpose is to achieve poverty reduction through economic growth, human resource development, women's status improvements, and environmental sustainability.[111] Access to modern energy services is a prerequisite for achieving these goals. The ABD achieves this mission by providing sovereign risk insurance for all projects in ADB member countries. The ADB also provides partial credit guarantees. In 2003, the ADB provided more than $6 billion in loans, as well as $179 million for technical development for projects and sectors in member countries.[112] The ADB lent $93 billion between 1968–2001, with about 20% allocated to the energy sector. It has focused primarily on transmission and distribution systems for electricity and natural gas with limited renewable energy financing.[113]

Risks of Renewable Energy Development

Today, renewable generation excluding large hydroelectric (microhydroelectric power, marine generation, solar and photovoltaic, wind power, low-temperature solar heating, geothermal heat, and biomass) accounts for approximately 2% of the world's energy generation. When large-scale hydropower projects, which have their own array of environmental issues, are included in the calculation, renewable resources account for 18% of worldwide generation.[114]

There have been limited increases in the financing of renewable energy projects. For example, an analysis of international financing institution (IFI) funding shows that, although overall international funding for energy projects decreased during the last decade, there has been a limited increase in funding of efficiency projects and renewable energy projects.[115] To develop a viable renewable energy power sector, a significant shift in institutional funding priorities is required. In fact, some have even recommended that over the next decade, IFIs engage in a massive funding shift, diverting funds to energy efficiency and renewables projects such as new biomass, microhydroelectric, diesel/wind hybrid, and solar systems in rural areas, in conjunction with a 10-year, worldwide moratorium on funding for coal, gas, and oil resources.[116]

Factors discouraging renewable investment

There are several factors that are discouraging investment in renewable energy projects. These include high up-front capital costs, the unavailability of renewable, energy-specific risk-mitigation instruments, the absence of governmental regulatory policies to encourage renewable energy development, and an industry perception of these projects as risky investments.

High up-front capital costs. The costs per kWh for each specific renewable energy technology are shown in table 10–2.

Table 10–2: Capital Costs of Renewable Energy Technologies[1]

Technology	Investment cost (US $) per kilowatt	Current cost per kilowatt hour	Estimated Future cost per kilowatt hour
Biomass Electricity Heat	500–6000 170–1000	3–12 ¢/kWh 1–6 ¢/kWh	4–10 ¢/kWh 1–5 ¢/kWh
Wind electricity	850–1700	4–8 ¢/kWh	3–10 ¢/kWh
Solar photovoltaic electricity	5000–18000	25–160 ¢/kWh	5–25 ¢/kWh
Solar thermal electricity	2500–6000	12–34 ¢/kWh	4–20 ¢/kWh
Micro Hydro	700–800	2–12 ¢/kWh	2–10 ¢/kWh
Geogthermal electricity	800–3000	2–10 ¢/kWh	1–8 ¢/kWh
Marine Energy Tidal Wave	1700–2500 2000–5000	8–15 ¢/kWh 10–30 ¢/kWh	8–15 ¢/kWh 5–10 ¢/kWh

[1] Adapted from World Energy Assessment Overview: 2004 Update, at 50.

Comparatively, up-front capital costs per kWh for large, on-grid, fossil-fuel, combined-cycle projects are in the range of 3 cents per kWh. Despite having higher initial costs, once they are operating, renewable projects have minimal generating costs. This allows a substantial portion of revenues to be devoted to loan repayment.[117] Furthermore, renewable projects are not exposed to fossil fuel price risks and carbon restrictions on fossil fuel projects.[118] A policy that encourages investment in innovative projects and pilot projects will increase experience with renewable generation and lead to new technologies for improving performance, each of which will lower generating costs.[119]

Renewable energy projects are considered riskier investments than conventional power generation. Another major hurdle in the development of renewable power is that financiers are averse to financing renewable energy projects and rate the risk of renewable energy technologies as higher than conventional energy generation technologies.[120] This results in increased expenses in the financing of renewable energy. As the perceived risks increase, financiers demand more project equity. This has proven problematic for renewable energy investors, as financiers have been demanding between 25% and 50% project equity, constraining project profitability and exposing investors to high-risk burdens.[121] Moreover, renewable power projects are not as financially attractive to investors because they generally have higher investment costs and lower projected rates of return than fossil fuel technologies.[122] The UNEP survey participants emphasized that financiers generally do not understand renewable energy technology and do not have experience financing these types of projects.

Stakeholders have expressed concern that the regulatory frameworks in developing countries are not sufficiently established to provide the regulatory and legal support needed for renewable energy project success. Stakeholders are concerned about the stability of existing regulatory frameworks.[123] Without regulatory stability, future revenues are uncertain, resulting in greater project risk and higher financing costs.

Renewable project development and power generation are not risk free. With geothermal projects, major risks include the long regulatory permitting process, drilling expenses, equipment costs, and exploration risk.[124] Risks inherent to photovoltaic technology include theft, component breakdown, and weather damage.[125] Different risks exist for wind power projects, whose project sponsors are concerned about the impact of lengthy permitting processes and long-term equipment capability.[126] For both solar and wind power, a concern or risk is supply intermittency. Similarly, project sponsors for biomass generation face concerns of resource availability, as well as resource pricing and emissions impacts. Hydropower projects are prone to seasonable flow variability, as well as environmental regulations. Wave/tidal power may be another viable renewable energy source, but technology is still in the development phase.[127]

Absence of renewable-energy-specific insurance instruments. Renewable energy projects, because of their small size, sometimes unproven technology, and high capital costs, are subject to unique project risks. There are no renewable-energy-specific risk-mitigation instruments to account for these unique risks. Developing such specific risk-mitigation instruments will be difficult because of the diversity of renewable energy technologies and regulatory systems employed by host governments. Respondents attributed this institutional inertia to unfamiliar and, in some instances, unproven renewable energy technologies, small project size, and perception of high risks in the renewable energy sector resulting from relative inexperience, which has discouraged the insurance of renewable energy projects. This has resulted in project sponsors attempting to use existing insurance instruments to insure renewable energy projects.

One means of resolving this paradox is to enhance exposure of renewable energy projects. Currently, insurance companies have limited data for determining coverage costs. Insurance institutions determine coverage costs by estimating the likelihood of certain event occurrences. With renewable energy projects, there is a lack of data to statistically access the likelihood event occurrence. This results in higher insurance costs. Only further operation and renewable energy sector establishment will provide these data.[128] This argues for a standard model for SPPs in developing nations, along the lines suggested in previous chapters.

A conclusion from the UNEP survey is that if commercial insurance were available to cover renewable-energy-specific risk mitigation, then investment in the renewable energy sector would quadruple.[129] However, such specific insurance products are currently not readily available. When renewable energy projects do obtain insurance, it is often modified versions of traditional insurance products sold at higher prices. Insurance products that have been adapted for renewable energy projects include construction all risks/erection all risks; delay in start-up/advance loss of profit; operating all risks/physical damage; machinery breakdown; business interruption; operator's extra expense; and general/third-party liability.[130] The availability of renewable-energy-specific risk-mitigation instruments is increasing. Such instruments include alternative risk-transfer products, specialist underwriting vehicles, weather derivatives, credit derivatives, and political risk insurance. According to UNEP research, weather insurance and credit derivatives are the most widely employed risk-mitigation instruments for renewable energy projects.[131]

Policy options to mitigate risk. Any renewables policy must signal market participants that supporting renewable generation is a long-term market objective of the host country.[132] A long-term enforceable PPA embodying a tariff scheme for a fixed term is the most important policy to enact for SPP development. Under this mechanism, a utility is required to buy all of the project's output at a predetermined price. Short-term tariffs do not have the same effect. For example, when tariffs on a microhydroelectricity project in Sri Lanka were set at short-run international oil costs, once the oil prices dropped, market development was constrained and investors encountered substantial losses.[133] (See chapter 11 for an analysis of PPAs.)

The foundation provided by a PPA can be augmented by a renewable power portfolio standard that obligates utilities to purchase, at market price, a certain percentage of power from renewable sources. Thus, this option requires the production of renewable sources to ensure the portfolio quota is achieved. When employed alone, however, a portfolio standard does not substitute for the requirement of an enforceable PPA at a set tariff. A pricing subsidy program, where the subsidy payment is the incremental cost of installing a renewable energy project (compared to a fossil fuel project) allows for more direct competition.[134] An alternative to subsidizing renewable projects is to reduce historical subsidies to fossil fuel generation. World Energy Assessment estimates that global subsidization of conventional energy is approximately $250 billion per year.[135] By requiring fossil fuel generators to internalize costs from the impact of their emissions and discharges, all projects would compete on equal footing.[136]

Next we will turn to the PPA, its role in SPP development, and an examination of key risk-mitigation provisions of the PPA.

Notes

1. Dubash, Navroz K., 2002. *Power politics: Equity and environment in electricity reform*. Washington, DC: World Resources Institute: 11–12.

2. Energy and Mining Sector Board (EMSB). 2001. *The World Bank Group's energy program: Poverty alleviation, sustainability, and selectivity*. Washington, DC: World Bank.

3. Dubash, The changing global context for electricity reform, 17. This reduced the funding and project guarantees traditionally available to public utilities. Moreover, the Bank's shift forced countries seeking to attract foreign investment to engage in market-based sector reforms.

4. EMSB. 2004. *Public and private sector roles in the supply of electricity services: Operational guidance for World Bank Group staff.* Washington, DC: World Bank.

5. Dubash, The changing global context for electricity reform, 16.

6. Dobozi, Istvan. 2003. Power purchase agreements and competitive power markets: Conflicts and reconciliation. Paper presented at the 2nd Energy Regulation and Investment Conference, Budapest, Hungary, May 8–9.

7. Dubash, *Power politics,* 1; Dubash, The changing global context for electricity reform, 11–12; Hunt, Sally. 2003. Competitive power markets? A viewpoint on the practical way forward for emerging and transitional economies. World Bank Energy Lecture Series, Washington DC, January 13.

8. Townsend, Alan. 2000. Energy access, energy demand, and the information deficit. *Energy services for the world's poor: Energy and development report 2000.* Washington, DC: World Bank and Energy Sector Management Assistance Programme.

9. Hunt, *Competitive power markets,* 7; EMSB. 2001. *Topical briefing to the board of directors on energy, the World Bank Group's energy program: Poverty alleviation, sustainability, and selectivity.* Washington, DC: World Bank.

10. Goldemberg, Jose, and Thomas B. Johansson, eds. 2004. *World energy assessment: Overview 2004 update.* Washington, DC: United Nations Development Programme.

11. Dubash, *Power politics,* 1.

12. EMSB, *Public and private sector roles in the supply of electricity services,* 2.

13. EMSB, *Topical briefing to the board of directors on energy*, 13.

14. Lamech, Ranjit, and Kazim Saeed. 2003. *What international investors look for when investing in developing countries: Results from a survey of international investors in the power sector.* Washington, DC: World Bank, Energy and Mining Sector Board; Goldemberg and Johansson, *World energy assessment*, 39.

15. EMSB, *Public and private sector roles in the supply of electricity services*, 2.

16. Dubash, *Power politics*, 4.

17. Matsukawa, Tomoko, Robert Sheppard, and Joseph Wright. 2003. Foreign exchange risk mitigation for power and water projects in developing countries. Discussion Paper No. 9, World Bank, Energy and Mining Sector Board; Lamech and Saeed, *What international investors look for when investing in developing countries*, 3.

18. EMSB, *Topical briefing to the board of directors on energy*, 14–16.

19. Fritsche, Uwe, and Felix Matthes. 2003. Changing course: A contribution to a global energy strategy. Paper no. 22, Heinrich Boll Foundation. Johannesburg, South Africa: Heinrich Boll Foundation.

20. International Energy Agency (IEA). 2002. *World energy outlook 2002.* Paris: IEA.

21. EMSB, *Public and private sector roles in the supply of electricity services*, 2.

22. United Nations Environmental Programme (UNEP), Division of Technology, Industry and Economics. 2004. *Financial risk management instruments for renewable energy projects: Summary document.*

23. Rowat, Malcolm. 1998. Financing international projects (Project finance and risk mitigation). Paper presented at the International Chamber of Commerce seminar. Paris, November 16.

24. Ibid.

25. UNEP, *Financial risk management instruments for renewable energy projects*, 14.

26. Ibid., 14.

27. Lamech and Saeed, *What international investors look for when investing in developing countries*, 4.

28. Ibid., 6.

29. Ibid., 8.

30. Ibid., 9.

31. Rowat, Financing international projects, 13.

32. Ibid., 8.

33. UNEP, *Financial risk management instruments for renewable energy projects*, 23.

34. Ibid., 9.

35. Ibid., 10.

36. Ibid., 30.

37. Smith, Robert. 2003. Mitigation expropriation risk for oil and gas investments in the Caspian region. Presentation to the IEA Roundtable on Caspian Oil and Gas Scenarios, April 15.

38. Ibid., 10.

39. Matsukawa, Sheppard, and Wright, Foreign exchange risk mitigation for power and water projects in developing countries, 1–2.

40. Ibid., 5.

41. Ibid., 1–2.

42. Matsukawa, Sheppard, and Wright, Foreign exchange risk mitigation for power and water projects in developing countries, 2–3; Gray, Philip, and Timothy Irwin. 2003. *Forex risk—A solution? Allocation exchange-rate risk in private-infrastructure contracts.* Washington, DC: World Bank.

43. Matsukawa, Sheppard, and Wright, Foreign exchange risk mitigation for power and water projects in developing countries, 2–3; Gray and Irwin, *Forex risk*, 3.

44. Gray and Irwin, *Forex risk*, 5–6.

45. Matsukawa, Sheppard, and Wright, Foreign exchange risk mitigation for power and water projects in developing countries, 3–4.

46. Gray and Irwin, *Forex risk*, 6.

47. Matsukawa, Sheppard, and Wright, Foreign exchange risk mitigation for power and water projects in developing countries, 3.

48. Gray and Irwin, *Forex risk*, 6.

49. Matsukawa, Sheppard, and Wright, Foreign exchange risk mitigation for power and water projects in developing countries, 6–8.

50. Rowat, Financing international projects, 34, 40.

51. Ibid., 26.

52. Matsukawa, Sheppard, and Wright, Foreign exchange risk mitigation for power and water projects in developing countries, 9.

53. Gray and Irwin, *Forex risk*, 2–3.

54. EMSB, *Public and private sector roles in the supply of electricity services*, 4.

55. World Bank. 2001. *California power crisis: Lessons for developing countries.* Washington, DC: World Bank.

56. EMSB, *Public and private sector roles in the supply of electricity services*, 4–5.

57. Ibid., 5.

58. World Bank Project Finance and Guarantee Department. 1997. *The World Bank guarantee: Catalyst for private capital flows, project finance and guarantees department, resource mobilization and cofinancing.* Washington, DC: World Bank.

59. Fritsche and Matthes, Changing course, 39.

60. Ibid., 39.

61. Ibid., 40, Table 4.

62. Rowat, Malcolm. 1998. Credit enhancements for private infrastructure financing. Presentation sponsored by LeBoeuf, Lamb, Greene & MacRae. San Francisco, CA, January 28.

63. World Bank Project Finance and Guarantee Department, *The World Bank guarantee*, 17.

64. EMSB, *Topical briefing to the board of directors on energy,* 22. The WBG will achieve its poverty-reduction objective by using energy financing to facilitate access to modern energy services, reduce the costs of energy supplied, ensure that subsidies reach the targeted population, and provide energy for improving social services. World Bank and IFC programs and private capital support are often employed to achieve these objectives.

65. Ibid., 23. Investments by the WBG can stabilize host country fiscal economies by attracting private capital to sector development. The host government can decrease direct budgetary support of the sector and divert these funds to other uses. In addition, WBG instruments can be employed to reduce the host government's exposure to energy-sector risks.

66. Ibid., 32. This includes investments relating to market regulatory mechanisms, competition expansion, asset divestiture, relaxing market-entry restrictions, and establishing local financial institutions. Loans from the IFC and World Bank and MIGA guarantees are financing instruments that are often used in these

situations. In conjunction with these instruments, the WBG often provides technical assistance to host governments regarding commercialization and privatization.

67. Ibid., 21. Employees of the WBG are advised to use energy-efficiency funds, rural energy funds, GEF funds, and the Prototype Carbon Fund to finance this objective.
68. Ibid., 18.
69. Ibid., 10.
70. Ibid., 17
71. Ibid., 19–20.
72. World Bank Group (WBG). 2003. *World Bank Group annual report.* www.worldbank.org/annualreport/2003/world_bank_group.html.
73. WBG. What is the World Bank. web.worldbank.org/WBSITE/EXTERNAL/EXTABOUTUS/0,,contentMDK:20040558~menuPK:34559~pagePK:34542~piPK:36600,00.html.
74. WBG, *World Bank Group annual report,* 50; International Bank for Reconstruction and Development (IBRD). 2003. Financial Statements and Internal Control Reports, June 30. Washington, DC: IBRD.
75. WBG, *World Bank Group annual report.*
76. Ibid.
77. Rowat, Financing international projects, 38.
78. Rowat, Credit enhancements for private infrastructure financing, 26.
79. World Bank Project Finance and Guarantee Department, *The World Bank guarantee,* 2.
80. Rowat, Financing international projects, 34, 40.
81. Ibid., 34.

82. Ibid., 28; World Bank Project Finance and Guarantee Department, *The World Bank guarantee*, 11.

83. WBN, *World Bank Group annual Report*.

84. World Bank Project Finance and Guarantee Department, *The World Bank guarantee*, 11.

85. WBG, What is the World Bank.

86. WBG, *World Bank Group annual report*.

87. Ibid.

88. Rowat, Financing international projects, 49.

89. WBG, *World Bank Group annual report*.

90. Ibid.

91. Fritsche and Matthes, Changing course, 42, Table 5.

92. Rowat, Financing international projects, 43–44.

93. Ibid., 45.

94. EMSB, *Topical briefing to the board of directors on energy*, 28.

95. Matsukawa, Sheppard, and Wright, Foreign exchange risk mitigation for power and water projects in developing countries, 9.

96. WBG, *World Bank Group annual report*.

97. Matsukawa, Sheppard, and Wright, Foreign exchange risk mitigation for power and water projects in developing countries, 10–11.

98. Fritsche and Matthes, Changing course, 57.

99. Prototype Carbon Fund (PCF). 2003. *Annual report 2003*. Washington, DC: Prototype Carbon Fund.

100. Ibid., 1.

101. Ibid., 2.

102. Miller, Alan S., and Eric Martinot. 2001. The Global Environment Facility: Financing and regulatory support for clean energy. *Natural Resources and the Environment* 15(3): 164–67.

103. Ibid., 2.

104. Ibid., 1.

105. Fritsche and Matthes, Changing course, 44.

106. Ibid.

107. WBG, *World Bank Group annual report*.

108. Fritsche and Matthes, Changing course, 49. Their share in the energy sector amounted to about 3.5% of total investments.

109. Fritsche and Matthes, Changing course, 51–52.

110. U.S. investors are lenders or investors domiciled in the United States that are more than 50% owned by U.S. citizens or foreign-domiciled corporations with more than 95% of ownership by U.S. citizens.

111. Asian Development Bank (ADB). 2005. Strategic development objectives. http://www.adb.org/About/objectives.asp.

112. ADB. 2005. Statement of ADB operations in 2003. http://www.adb.org/Documents/reports/operations/2003/default.asp.

113. Fritsche and Matthes, Changing courses, 45.

114. Goldemberg and Johansson, *World energy assessment*, 48–49.

115. Fritsche, Uwe R. 2004. Financing renewables: A core issue of the global energy strategy, and of the Renewables 2004 Conference. Presentation to the WRI/Heinrich Boll Foundation/IIEC Conference, Washington DC, April 21.

116. Fritsche, Financing renewables, 6–7.

117. Papathanasiou, Demetrios. 2004. Policy options to promote renewable energy in China. Unpublished manuscript. Policy paper for the World Bank, draft for comment. March 6.

118. Renewables 2004: International Conference for Renewable Energies, Conference Issue Paper. Sponsored by Heinrich Boll Foundation. June 1–4, 2004, Bonn, Germany.

119. Papathanasiou, Policy options to promote renewable energy in China, 5–6.

120. Renewables 2004 Conference Paper, 11.

121. UNEP, *Financial risk management instruments for renewable energy projects*, 14.

122. Renewables 2004 Conference Paper, 17.

123. UNEP, *Financial risk management instruments for renewable energy projects*, 16–17.

124. Ibid., 18.

125. Ibid., 18.

126. Ibid., 18.

127. Ibid., 18.

128. Ibid., 16–20.

129. Ibid., 19.

130. Ibid., 21–22.

131. Ibid., 29.

132. Papathanasiou, Policy options to promote renewable energy in China, 5–6.

133. Miller and Martinot, The Global Environmental Facility, 6.

134. Papathanasiou, Policy options to promote renewable energy in China, 3, 14.

135. Goldemberg and Johansson, *World energy assessment*, 72.

136. Ibid., 13.

11 Key Provisions in Power Purchase Agreements

Overview

The power purchase agreement (PPA) is the key prerequisite for attracting private capital to the power sector. But the sector structure also must be in place. To successfully attract significant private capital to the electricity sector, several elements must be present:

- There must be a system of government regulation for the siting and permitting of power facilities that is explicit, substantive, and procedurally accessible. A single permit administered by one ministry is preferable for attracting investment.

- The rights of private independent power developers must be established and vested by contract. The contract must vest rights to develop, build, and operate a power-generating facility, with a guaranteed long-term contractual sale through interconnection to a stable purchaser of the power. The payments by that purchaser must be secure for the life of the power sale contract. This may require a government guarantee of the utility obligation, opportunities for third-party direct retail sales or export of power, or special credits for renewable power development. In the five Asian nations examined, one of the primary concerns of stakeholders was that some utilities had not paid full avoided cost (Sri Lanka and Indonesia). Experience in Indonesia and India also demonstrated that independent power producer (IPP) contract prices would not necessarily be respected when currency devaluations altered the expected cost of the power in equivalent local currency.

- The price of power purchased must be contractually established for the duration of the term of the contract, with appropriate adjustment clauses (escalators) to account for changes in the costs of operation attributable to changes in fuel costs, operation and maintenance, taxes, laws or regulations, and so forth. The price component must sufficiently cover debt repayment for the life of the project debt under all possible scenarios under the key contracts, as well as provide a sufficient residual to compensate the equity investment and risk. A primary stakeholder concern in the Asian countries surveyed was the security and adequacy of the tariff price protection over time. If this protection is provided securely in the PPA, it provides credit support for long-term financing of the SPP project. Not all these countries provide this, although all these countries have a policy of adjusting the tariff periodically to reflect changes in avoided energy costs. However, because the tariff level from year to year is not contractually guaranteed or locked into an objective formula in most of these programs, it does not provide bankable credit support for project financing.

- For the private developer, fuel supply and transportation, equipment acquisition, construction, and operation and management all must be established securely by contract, within the context of applicable laws and regulations. These contracts must offer sufficient credit support and guarantees to ensure project viability and timely completion at fixed costs. A significant stakeholder concern in each of the five Asian countries examined is the practice of the ministry and utility altering the contract price or the contract tariff-escalation factor in contravention of the original program design or contract requirements. Indonesia and some Indian states have balked at enforcing contract provisions in IPP contracts.

- A lender's right of notice of deficiencies and right to cure breaches or defaults in the duties and obligations owed by the independent power facility must be included in the power sale contract to prevent loss of the long-term benefits of power sale. This is necessary in any project financed by international capital. For smaller renewable energy projects financed by local equity and debt capital, these requirements may not be as exigent.

- There needs to be recourse to a system of law that provides speedy and impartial resolution of disputes that cannot be mutually resolved. In most instances, this requires recourse to a proven, swift, and procedurally guaranteed right to recourse in a court in a stable country with a neutral, independent judiciary that honors the rights of timely appeal, as well as the ability to enforce a judgment in the host country. Experiences in some developing countries have proved that expectations of such neutral judicial recourse have not, in fact, been available. By contrast, the choice of forum for dispute resolution and choice of law to apply can be elected. Larger IPP contracts may designate international tribunals or choice of law to remove adjudications from the bias of a host country or to seek a more neutral forum. Some developing nations have not respected such choices or have refused to enforce judgments rendered by courts or international tribunals.

- To attract international capital, the contract must contain a mechanism that adjusts for fluctuation in foreign exchange ratios so that project cash flow is held constant in the converted currency (or currencies) in which the international investments in capital equipment are denominated or an internationally acceptable currency. There must be sufficient availability of these foreign exchange currencies. Many of the smaller renewable power projects surveyed here are not dependent on a substantial foreign-exchange-denominated component or foreign lending, as are larger IPP projects. Most of the SPP programs do not adjust for foreign exchange, on the theory that borrowing will be local. The Indonesian SPP program did attempt to adjust capacity payments for changes in the cost of capacity, with a substantial offshore capital component.

The PPA is the foundation on which successful SPP projects proceed. This contract is the legal prerequisite for financing an SPP. It is the skin of the operating relationship of the parties and controls these relationships for decades. As a binding legal relationship, it is essential that the contract anticipate a variety of construction and operating contingencies and establish operating parameters that govern the sale of the invisible, unable to be stored electric commodity moving at the speed of light.

The importance of a standardized, properly designed PPA in a developing country is that it provides a fair, neutral, and financeable contractual arrangement between the power seller and purchaser. The PPA embodies the core legal relationship for subsequent third-party contracts and credit relationships with lenders, equity participants, and equipment suppliers. With independent private power development, the contract becomes the legal vehicle for the transaction in electric power.

Ultimately, the PPA must meet the norms required for debt financing of the project and the attraction of equity interests and capital. Because of this necessity, the PPA must be constructed around certain legal provisions to provide security for equity interests, lenders, and investors. Lenders evaluate the project's economics, its prospect for successful permitting, and the PPA as the legal embodiment of the lender's rights and protections. What has evolved in SPP project finance is the expectation that the PPA will satisfy a set of basic requirements. This translates into a conventional PPA format that has evolved from the U.S. experience with the Public Utilities Regulatory Policies Act (PURPA) to many international locales.[1] If it does not contain these elements, the project is unable to be financed and typically fails. The applicable law for resolving disputes is analyzed in chapter 12.

The tariff embedded in the PPA establishes the economic framework for the long-term power transaction. Tariffs may need to change over time to reflect the fluctuating marginal cost of power production for the utility. Although the determination of a proper tariff is a complex and sophisticated undertaking that typically requires analysis by experts, its impact is straightforward for an individual prospective SPP: At the tariff provided by contract, the transaction either does or does not work financially. A pro forma analysis determines whether the tariff is sufficient to motivate SPP development.

The PPAs in the five Asian nations were profiled in chapters 3–9. These PPAs, by design, are intended to be simple, straightforward, and transparent. This chapter discusses typical PPA provisions. It analyzes the range of provisions in PPAs for simple SPPs and more complex IPPs. Therefore, the scope of the analysis that follows includes complex PPA clauses that may be included in certain IPP contracts and individually negotiated SPP agreements, but may not always be included in a simplified, standardized PPA. Highlighted throughout this analysis are "typical" provisions, the most liberal and conservative variations of these typical provisions, and commentary on how these provisions should be drafted and function.

Contract Formation and Contract Validity

Traditionally, to make contracts enforceable, there must be an offer and acceptance supported by consideration.[2] The law also requires that the parties specify their bargain in terms that are sufficiently precise to permit a court to determine an appropriate remedy for breach.[3] Courts can look past the contract and use the parties' course of dealings to define contract terms.

Safe and reliable project operation

Typically, a provision of the power sale contract sets forth the standards for facility operation. It ensures the buyer of power that the generating facility will be operated to a standard that assures continued power output. It also states the standards the seller must satisfy to prevent breach. The standards are typically very general, such as "good utilities practice" or state utility industry standards. A typical provision addresses a general standard and references a definition that includes a more specific set of standards, such as the National Electrical Safety Code, the National Electric Code, or other government codes. The contract might require the facility to be maintained in accordance with Prudent Electrical Practice under the safe and reliable operation provision. Prudent Electrical Practice is typically defined as "those practices, methods and equipment as changed from time to time, that are commonly used in prudent electrical engineering and operations to operate electric equipment lawfully and with safety, and dependability and that are in accordance with the National Electrical Safety Code, the National Electrical Code or any other applicable Federal, State and Local government codes." A contract may define Good Utility Practice as "the practices, methods and acts engaged in or approved by a significant portion of the electric utility industry." Additional provisions accompanying these standards include notice to the other party of maintenance outages and requiring alterations in operation if the buyer's system is modified.

For example, a PPA could provide, "[i]f future alterations to the interconnecting circuit, including but not limited to conversion to a higher voltage, are required, the seller shall be responsible for all interconnection changes necessitated by the alteration and shall bear all costs associated with these changes." Or a provision may provide the buyer the "right to monitor operations of the Project and may require changes in Seller's method of operation if such changes are necessary, in Buyer's sole judgment, to maintain Buyer's electric system integrity."

The least specific PPAs mention "good utility practices" without defining the term. However, that same PPA may set forth in great detail certain metering and interconnection provisions that govern the technical equipment and operations for power transfer. In one contract reviewed, a maintenance standard was not mentioned. When a PPA is vague, it forces the parties to rely on common law and state law to remedy a breach.

Stringent contract provisions may require the maintenance of operational records in addition to the common guidelines for operation. A stringent procedure might require, "Seller shall keep and make available adequate maintenance logs…perform any necessary unscheduled maintenance on an emergency basis" and "if Buyer is not satisfied with the schedule proposed by Seller, Buyer will specify the maintenance schedule to be followed." Monthly reports may specify to the minute how quickly status questions should be answered, identify a required amount of "spinning" reserves, specify when the generators can go offline, specify staffing levels at the project control room,[4] or require an independent review of calibration results and maintenance procedures.

Incentives for better performance

Some PPAs contain a payment structure that provides incentives for the independent producer to achieve greater availability or capacity. These are designed to use financial incentives rather than coercion to cause the SPP to maximize availability—in most developing nations, there is a shortage of capacity to satisfy pent-up demand. These incentives are designed to encourage the SPP to be available at peak daily or seasonal times to help satisfy system peak demand.

Many PPAs, particularly for larger IPPs, contain performance criteria requiring the plant to be available a certain amount of time and at an average efficiency during those times. In general, the utility's obligation to pay is conditional on the independent producer's satisfaction of the performance criteria. If the independent producer does not meet the required performance levels, it does not receive full price for its energy sale.

Performance criteria usually consist of two elements: (1) performance standards for the plant's availability and (2) periodic testing of the plant's capacity.[5] Performance standards may require the plant to be available a certain percentage of time during a peak season or year. Periodic testing requires the plant to meet certain capacity ratings to ensure that its equipment is adequately maintained. Power generation equipment can degrade over time, and availability and capacity will decline.[6] Therefore, many contracts have a "dead band" of tolerance of deviation that accounts for normal capacity degradation over time.

In some of the countries, such as Sri Lanka, the utility pays only for the energy delivered, without explicitly paying for the capacity value provided by the SPPs, which are hydroelectric projects. Although some intermittent renewable energy projects cannot provide reliable demand, cogeneration, landfill-gas, many hydroelectric, and biomass projects can reliably provide demand to the utility. Generally, when the SPP does supply reliable demand, it should be entitled to a tariff component that reflects this contribution. When it does not, it typically does not receive a demand component in the tariff.

In situations where an energy-only SPP delivers reliably, an incentive payment is appropriate if the intermittent SPP does supply at key times. The Indonesian program does this by steeply increasing the on-peak tariff compared to the off-peak tariff. In this way, the Indonesian program design replaces cumbersome, coercive PPA provisions with financial and market incentives. Some contracts provide for a direct cash bonus as part of the tariff payment, whereas others allow a carryover of surplus or deficit availability value to the following year.[7]

In the converse situation, where a demand payment is included in the PPA tariff, a penalty clause to reduce this payment may be appropriate when the SPP repeatedly fails to deliver. Some of the countries profiled earlier incorporate this. Typically, PPAs reduce tariff payments by a certain percentage as a penalty for failing to meet the availability or capacity target levels, but others do so only for each day of forced outage in excess of the permitted number of days. Some contracts require the developer to compensate the utility for having to procure supply from another source. In some instances, if the developer continuously fails to meet availability or capacity levels, the utility may have the right to terminate the contract or permanently reduce payment.[8]

Metering power sales

The metering provision addresses the measurement of the billable commodity, electric power. A metering device is installed near the buyer's facilities to measure the amount of commodity received through the meter. Although there is no clear common law principle addressing the accuracy of meter reading and billing, the principles of good faith and fair dealing should apply.[9] A common PPA metering provision provides for the metering facilities owned, operated, and maintained by the buyer at the seller's expense. Meters typically are read monthly and tested annually by the buyer at the seller's expense. When there is not a simultaneous transfer of power, line losses typically are deducted from the billing to the buyer's advantage. A typical provision might read, "electricity delivered by Seller to Buyer will be measured by electric meters and associated equipment of a type approved by the Commission, which meters and equipment shall be installed on Seller's premises and owned, installed, operated and maintained by Buyer at Seller's expense. Seller shall, at its own expense, furnish, install and maintain mounting facilities for such meters and associated devices."

If tests show the meters to be out of calibration by more than the typical 0.5%–2%, past billings are corrected retroactively by assuming the meter failed at the midpoint since the last test. A corrected billing is extrapolated from this presumption. This correction is usually retroactive to one-half the time since the last inspection, with allowance for verifiable adjustments. The retroactivity usually has an expressed or implied limit of six months.

In addition to these common terms, some contracts include terms such as the installation of a second set of meters, stricter standards for meter accuracy, and special cost responsibilities for meter testing.

Some PPAs address the possibility of and payment for additional accuracy tests done outside the normal plan. When one party commands a meter test and discovers an inaccurate meter, the cost of that inspection typically is shifted to the other party. This cost shifting appears to be an effective incentive for both parties to carefully control meter accuracy. A unique term in one contract surveyed allowed an independent review of the calibration tests. Failure to provide a specific provision could make it unclear whether a small error reading is allowable and whether this error should be carried into the billings or addressed retroactively.

Interconnection to the grid

This provision is typically straightforward and clarifies which party is responsible for the facilities tying the independent power provider to the power network. Because the seller is tapping into an existing network of power transmission and distribution, it is logical that it should bear the cost of modifying those facilities. The most general and typical provisions in PPAs include the facilities built, owned, and operated by the buyer at the seller's expense. The seller's expense is usually determined as a monthly payment that is updated periodically to reflect actual expenses. A typical provision might read, "Company will, if necessary, and at Cogenerator's expense, also modify its existing switching station and transmission facilities and construct new transmission and/or distribution facilities… Cogenerator will, at its expense, construct the Cogenerator Interconnection Facilities, being the facilities interconnecting the Cogenerator Substation with the Interconnection Facilities…Cogenerator will pay to Company monthly…the monthly amount specified…shall be adjusted…to reflect increases" of operation.

Variables in this provision usually specify which party designs and pays for the facilities, which may be the responsibility of either party. Some strict PPA provisions include terms that force the seller to change interconnection facilities to accommodate system modifications of the buyer. Some are specific enough to require periodic tests of the interconnection facilities. There are multiple options for interconnection agreements. One option gives the responsibility for the design, purchase, construction, ownership, and operation to the buyer at the seller's expense. A second option allows the seller to either install some of the equipment or pay for installation before the start of construction by the buyer. The final option gives all responsibilities to the seller; the buyer keeps the right to review the design for adequacy. All of these options reserve a great deal of power in the buyer, even with the option allowing seller ownership.

Milestones for project development

Generally, milestones are used to address key dates, achievements, or acts that the parties agree are events of preliminary default or that trigger a modification of terms. A typical milestone provision includes dates that must be satisfied for financing, environmental permits, start of construction, completion of construction, online operation, and periodic project status reviews. If these milestones are not met, the contract may be terminated or the nonfulfilling party may be considered in breach. Less common elements of milestone provisions include periodic reviews, notice prior to power delivery, identification of a project manager, regulatory commission approval, financing verification, quarterly project reviews, and forfeiture of IPP deposits if milestones are not met.

It is questionable whether the common law doctrine of anticipatory repudiation would cover the failure to achieve typical milestone terms. Anticipatory repudiation requires an actual, unequivocal manifestation of intent not to perform.[10] These milestone provisions become effective before an independent anticipatory repudiation, regardless of whether that milestone is material to total contract performance.

Utility acquisition of the project and first refusal

The PPA may allocate certain additional rights to the utility, such as the right of first refusal to extend the PPA or purchase the facility, emissions credits, or limits on the location of the facility. Emissions credits, such as those offsetting CO_2 emissions, may be allocated to the buyer. A PPA grant of the right of first refusal provides the utility the option to purchase the facility, capacity, or energy after the termination of the PPA. When the plan is to transfer the facility to the utility once it is depreciated, the PPA may grant a right of first refusal on the purchase of the seller's facility for consideration at the expiration of the contract term. Most clauses require the utility to

Key Provisions in Power Purchase Agreements

meet the offer of a bone fide third party, whereas others allow the utility to purchase the interest at its fair market value, as agreed to by the parties or determined by an independent appraiser or neutral methodology.

Although the fair market value of a particular facility is difficult to estimate, it is the recommended standard for the right of first refusal. Potential third-party purchasers may not undertake the time and expense of evaluating the facility if they know the utility can step in and buy it at less than fair market value. Rights of first refusal should not be open ended. The parties must limit how, when, and under what circumstances the utility can exercise this right.

Modification of the Contract

Once the contract is executed, how can it be changed? Generally, if the parties have the capacity to make a contract, they also have the capacity to modify it.[11] This capacity is typically limited in the contracts by negating changes in duties except through written modification, waiver, or assignment of rights.

Modification of key provisions

This provision addresses whether the written contract can be modified and, if so, within what parameters. A typical modification provision simply states that modifications to the contract must be made in writing, executed by both parties. A typical provision might read, "No modification to this Agreement shall be binding on either party unless it shall be in writing and signed by both parties." The typical provision does not take away from or add to any common law rights. If the parties agree to change the contract provisions after the contract is formed, these modifications need only be agreed to by the parties.[12]

Other terms that may be included are the exclusion of certain modifications, designation of whether the modification of one provision excludes the modification of others, control of whether facilities can be modified, and specification of whether waivers of PPA rights require regulatory approval. Some contracts do not address modification, but do allow for written waiver. Some PPAs provide a "most favored nation"

257

provision, which gives the parties the right to more favorable exchange terms if one party negotiates better rates with a third party. The terms of this provision might be, "if Company (Buyer) believes, in its sole judgment, that any of such Other Agreements contains price terms, contract deposit terms, or security terms more favorable to the applicable third party purchaser than the terms of this Agreement are to Company, then Seller shall make such more favorable terms available to Company for a period of thirty (30) days after the date on which such Other Agreement was provided to Company."

Fuel-adjustment clauses

Some PPAs, especially for fossil fuel-fired IPPs and SPPs, contain fuel-adjustment clauses. Fuel is the major cost of operating a power plant. Retail utilities typically can pass on increases in fuel costs to ratepayers under an automatic fuel-adjustment clause or by relief from the regulatory commission. Independent power producers cannot do so without a fuel-adjustment clause in the PPA; without such a clause, the project must absorb the difference in fuel price changes.

Fuel-adjustment clauses allow independent producers to pass on some of the increased cost to utilities—and ultimately to their retail customers and ratepayers—by linking the price of energy to variable fuel costs. This is typically accomplished by a clause in the PPA that links the SPP or IPP power sale tariff to a national index (which can be weighted by various types, or a weighted "basket" of fossil fuels can be indexed to foreign exchange risk when that fuel must be imported and reflects transportation costs of the fuel delivery) or by changes in the utility's own cost of fuel. The former approach makes the fuel adjustment sensitive to the SPP project's proxy operating costs. The latter approach reflects the purchasing utility's avoided operating costs of fuel.

It is important to note that with most renewable energy projects, fossil fuel price fluctuations are not as critical. Wind and solar power flows are free. Unless there is a variable charge for use of the hydropower resource or for the use and combustion of landfill gas or other agricultural or biomass wastes, typically there is no cost for the fuel for these renewable resources. With trash-to-energy projects, the project developer or its waste-transfer agent is actually paid to take the fuel source, in lieu of the waste generator paying to place the waste in a landfill.

Many SPP projects, such as those discussed in Indonesia, India, and Thailand, include fossil-fueled, as well as biomass-fueled, cogeneration as eligible projects. In such instances, these small cogeneration facilities are very sensitive to fossil fuel price and availability changes. It is necessary to include an adjustment mechanism for these projects in the PPA, given the volatile nature of the fossil fuel commodity markets.

In structuring a renewable PPA, this does not necessarily mean that fuel price adjustment has no place. As discussed in chapter 2, renewable energy projects benefit the electric diversity of a nation by relieving the loading of the transmission and distribution infrastructure and by increasing efficiency. If larger fossil fuel-fired IPPs benefit from fuel-adjustment clauses in their PPAs so as to give these projects tariff protection, facilitate financing, and protect return on project equity, why shouldn't smaller renewable SPPs benefit from such protections?

As discussed earlier,[13] the Indonesian project design put a floor under the PPA tariff and allowed the tariff to increase with increases in the regionally disaggregated price of delivered oil, the marginal fuel for the state utility system. Such creative contractual provisions allocate value to the renewable project PPA in return for the benefits that the project provides the utility system and national energy posture. They also inject a dynamic component into the PPA that ensures that larger fossil fuel tariffs will not outpace renewable SPP tariffs if fossil fuel prices escalate over the term of the contract, as many observers expect will occur.

However, these clauses do not eliminate all fuel price risk. The SPP or IPP still must manage its fuel supply—for example, by entering into long-term contracts for its supply or by obtaining its supply from diverse sources.[14] In general, the independent power plant must take reasonable steps to control fuel price risk, but with an adjustment clause, which escalates when fossil fuel prices increase, it can pass the remaining risk on to the utility purchaser and its ratepayers. Because the utility would be running its marginal fossil fuel-fired thermal units to supply power if the SPP were not available—and so would incur these oscillating fossil fuel prices itself—if the PPA fuel-adjustment clause is properly structured, the utility and its customers should be indifferent as to whether the higher fossil fuel prices are paid by the utility to operate its own marginal fossil units or to the SPP to run its project and displace the least efficient utility plants.

Waiver of contract rights

The waiver provisions within contracts attempt to limit the application of common law waiver law. At common law, waiver is the relinquishment of a contract right through failure to require performance from the other party.[15] Waiver does not require mutual assent or consideration, and waivers are not otherwise subject to any requirement of a writing.[16] Oral waiver can be made even if the contract requires modifications to be made in writing.[17]

Most PPAs require modifications to be made in writing and include at least general waiver statements. The typical waiver provision allows waivers only in writing and tries to protect the parties in the event one fails to insist on strict performance of a particular duty or if one fails to assert one's rights under the contract. A typical PPA waiver provision might state, "The failure of either party to require compliance with any provision of this Agreement shall not affect that Party's right to later enforce the same. It is agreed that the waiver by either Party or performance of any of the terms of this Agreement or of any breach thereof will not be held or deemed to be a waiver by that Party of any subsequent failure to perform the same or any other term or condition of this Agreement or of any breach thereof."

Additional terms included in some contracts provide for the severability of other provisions not waived and waiver or no waiver of rights resulting from new laws that become effective after the contract execution. Although some PPAs include no waiver provisions at all, leaving recourse to common law waiver, the strictest PPAs include general waiver language and survival of the other contract provisions if one provision is waived. A typical provision might read, "If any clause, provision, or section of this Agreement be ruled invalid by any court of competent jurisdiction, the invalidity of such clause, provision, or sections shall not affect any of the remaining provisions."

Assignment of rights

These provisions address whether the contract rights and duties can be transferred from one of the original contract parties to a third party. Essentially, this means that the independent producer cannot assign its interest under the contract to a third party without the consent of the utility. These clauses ensure that the project owners continue to meet their contractual obligations in the event of changes in ownership, financing, or corporate structure. Although these clauses are designed to ensure that third-party assignments will not affect the delivery of power to the utility, they also constrain efficient ownership, financing, and project structural changes. The utility's desire to control assignment must not conflict with the rights of the project financiers to take contingent assignment of the project's physical and contractual assets to secure their loans to the project. Ideally, these clauses would require only consent to assignments that are not considered routine.

Under common law, rights are generally assignable, but duties may or may not be assignable.[18] Duties may be delegated, but the delegate ultimately remains responsible for performance. When contract provisions try to circumvent this rule by specifying that the "contract will not be assigned," courts typically have interpreted this as a prohibition against the delegation of duties, but not the assignment of rights.[19]

Typically there is very little deviation within PPAs. A typical assignment provision requires assignment by written consent of the nonassigning party, whose approval cannot be unreasonably withheld. A typical PPA provision might state that "nether party shall assign, pledge or otherwise transfer this Agreement or any right or obligation under this Agreement without first obtaining the other party's written consent, which consent shall not be unreasonably withheld." The assignee usually assumes both duties and rights. Additional variables in some PPAs include an ability to assign to affiliates without consent, prohibitions on assignment, and stipulations that the buyer of the contract must approve certain construction milestones and terms.

Dispute Resolution

Dispute resolution clauses attempt to keep parties out of lengthy and costly litigation. The typical dispute resolution provisions include an arbitration provision, in addition to specific provisions outlining breach. Arbitration typically can be accomplished more expeditiously and at lesser cost than awaiting the often slow pace of litigation. Arbitration can be controversial because of the perceived penchant of many arbitrators to move toward a middle ground rather than a decisive judgment in favor of one party to the dispute.

An arbitration provision states how disputes over contract performance or contract terms should be resolved. About half of the PPAs employed in the power sectors for SPPs include arbitration provisions. The typical provision provides that an arbitrator will be selected by both parties or, if they cannot agree, multiple arbitrators will be selected or an impartial party will select one individual. For example, a typical provision might provide that if a dispute arises, either party can give notice of its desire to submit the dispute to an arbitrator. The parties mutually select a single arbitrator, or, if they disagree on this issue also, an arbitration association or group will be asked to appoint an arbitrator. This arbitrator is given full authority to act pursuant to the contract. A U.S. provision may refer disputes to the American Arbitration Association or send issues to arbitrators in accord with a state arbitration act. Typically, the decision of the arbitrator is final, short of requesting a competent court to review possible errors of law.

Breach of the Contract

In bilateral PPAs, each party makes promises to the other. The seller usually promises to deliver power, and the buyer promises to take and pay for the power. These promises are the basis of actions for breach of contract if the promises are not fulfilled.[20]

Excuse, force majeure, and impracticability

These provisions attempt to excuse an event triggering default when that event is uncontrollable and temporary, or in some instances, permanent. Generally, at law, obligations may be excused if a situation occurs that makes performance impossible or highly impracticable.[21] However, to meet this general common law standard to excuse performance in the absence of a specific force majeure provision, the party asserting the impracticability defense must prove that the event was (1) not foreseeable, (2) not within its control to prevent, and (3) could not have been avoided by other risk-mitigation techniques (in other words, the asserting party did not assume the risk of the event). At the very least, this often is contentious and results in a suspension of power purchase and production until the dispute can be resolved through a lengthy and sometimes costly dispute resolution process.

It is highly preferable to include a force majeure clause in the PPA, specifically articulating which events constitute force majeure and precisely how the obligations of the parties are affected and will be suspended or excused in such events. These clauses set forth, in advance, how the parties agree to excuse the duty to perform when certain events occur that are not within the reasonable control of the parties and are not caused by their fault or negligence. A typical provision addresses unforeseeable causes beyond the reasonable control of the affected party and excludes labor disputes, nonavailability of fuel,[22] and actions, negligence, and willful misconduct of the party claiming force majeure.

Events included in these provisions can be grouped in two general categories: acts of God and certain controllable acts. Acts of God that may excuse performance typically include droughts, floods, earthquakes, storms, fires, unforeseeable accidents, poor weather, and other natural disasters. Other unforeseen events may include war, riot, and civil disobedience. Controllable acts include labor problems, such as strikes and walkouts, or material supply problems.

The most liberal provisions use general language that includes controllable acts and acts of God. For example, a PPA provision might provide that force majeure shall include "without limitations, failure or interruption of services due to causes beyond a party's control, sabotage, strikes, acts of God, drought or accidents not reasonable." Some such clauses broadly include "loss or impairment of the supply of electricity…failure or improper operation of transformers, switches or the other equipment necessary for receipt of electric energy…failure of any major supplier to perform" under the definition of Force Majeure.

The strictest provisions require notice of the force majeure, reasonable measures undertaken by the affected party to continue to perform, attempts to cure the problem, efforts to limit damages, as well as exclusion of excuse for problems originating before the force majeure. A provision may excuse performance because of force majeure "provided that the non-performing Party shall: (1) Provide prompt notice to the other Party of the occurrence of the Force Majeure giving an estimation of its expected duration and the probable impact on the performance of its obligations hereunder and submitting good and satisfactory evidence of the Force Majeure; (2) exercise all reasonable efforts to continue to perform its obligations hereunder; (3) expeditiously take action to correct or cure...; (4) exercise all reasonable efforts to mitigate or limit damages to the other party...; and provided further, that any obligation of either Party which arose before the occurrence of the Force Majeure event...shall not be excused." Some of the stricter provisions also exclude failure to make payments because of force majeure. A PPA may state, "Any provision of this Agreement requiring payment of money, to each of which Force Majeure specifically does not apply."

Unique provisions found in some PPAs excuse performance for specific upstream problems: failed generation machinery, loss of power or fuel to operate the facility, failure of the equipment necessary to transfer or receive electric energy, and delay in delivery of components for building the facility. One such provision states, "Uncontrollable Forces shall also include delays in receipt of generator rotor, generator stator, main power transformer or steam turbine caused by damage or loss in shipping." Another includes "defects in the design or manufacture of the gas turbine generator rotor and stator, steam turbine rotor, steam turbine generator rotor and stator, and main transformers of the Plant (but only where such defect was not reasonably discoverable" as force majeure.

Generally, force majeure clauses differ only in their inclusion or exclusion of certain triggering events.[33] At a minimum, force majeure provisions should contain all events that are reasonably beyond the control of the independent SPP developer.[34] Most force majeure clauses delay or cancel the party's performance depending on the nature and duration of the event. Generally, the party claiming force majeure is excused from performance without fear of contract termination, breach of contract, or other liabilities for nonperformance.[25] These clauses provide specificity to the already available, but difficult to prove, common law impracticability defense. This allows precision and preagreement as to the choreography of what transpires when uncontrollable events occur.

Change of law and "regulatory out" clauses

Some power contracts protect both parties—but particularly the SPP or IPP—from unanticipated changes in law and regulation. These clauses often are referred to as changed circumstances, change-in-law, or contract "reopener" provisions. Utility-owned power generation projects typically pass along the financial impact of regulatory changes to their ratepayers. However, independent producers cannot do the same without a provision to reopen or adjust the tariff for the increased costs resulting from these changes. Provisions must be expressly included in the PPA to allow for such adjustments. If they are not included, a dispute typically results between the power producer or seller and the power buyer as to whether these changes rise to the level of a force majeure event.

When the SPP asserts that it is disadvantaged by a change in law, it inherits the burden of proving that this change rises to a force majeure event in the absence of a change-of-law provision in the PPA. Recent experience suggests that the SPP, rather than the purchasing utility, is more likely to seek redress during the contract term for a change of law.

- The utility is the stronger party and the "home team," and therefore legislative changes are less likely to disadvantage the national utility.

- Some of the changes in law are changes in standards, operating procedures, or requirements that are unilaterally imposed by the utility and affect the SPP through a PPA clause that the SPP will comply with all regulations, procedures, and standards.

- When the utility buyer is disadvantaged by circumstances, it may cause the law or regulation to be changed to rectify the disadvantage, without recourse to a change-in-law provision in the PPA.

- During the Asian financial crisis, the Indonesian state utility, backed by the government and the local courts, refused to comply with the PPA payment obligations to IPPs; where there is self-help in this fashion, the PPA provisions are largely ignored.

Change-of-law clauses in PPAs may work in several ways:

- Automatic specified adjustment, as for the costs of compliance with a new regulation

- Requirement to renegotiate a PPA provision in good faith among the parties

- Alterations in the PPA to be determined by an independent neutral expert

- Indexation of payments to foreign exchange, fuel, or other indexes
- Percentage sharing among buyer and seller of power for additional costs

Change-of-law provisions usually include changes in environmental regulation or tax code treatment. In some situations, independent producers may be grandfathered or exempted from meeting certain changes in law; nevertheless, the parties should provide a provision for reopening the terms of the contract or otherwise making equitable adjustments to reflect a fair sharing of risks, if necessary. Without change-of-law provisions, international lenders will not advance the debt capital necessary for project finance.

Some PPAs contain "regulatory out" clauses. These clauses relieve the utility from some or all of its contract payments if a regulatory commission or other government agency prohibits it from passing its full SPP power purchase costs on to its ratepayers. This situation rarely occurs in a mature regulatory system: A regulatory commission only disallows or disapproves of cost recovery if the utility has acted with imprudence. Without a regulatory out clause, the cash flow of the entire project can be at risk because the independent producer depends on the utility's payments to cover operating expenses. In some instances, investors have refused to finance contracts that contain very broad regulatory out clauses, such as those that provide for an immediate reduction in energy and capacity payments or allow the utility to terminate the contract at any time if it is in its ratepayers' "best interest."[26]

Termination of contract

Generally, because expressed contract terms must be complied with literally, the terms of these contacts must be satisfied exactly as described, or breach will result.[27] A typical termination provision in these contracts allows only bankruptcy or missed milestones to trigger termination after the injured party gives notice. A typical provision may allow termination after an event of default by giving 30 days written notice to the other party. Additional triggers may be included in the events that cause breach. Default or breach typically is addressed as part of the termination section of the contracts. Default and breach terms are usually the same as those defined for termination of the contract: failure to make payment, failure to take electricity, missed milestones, or bankruptcy.

Miscellaneous provisions

Clauses that provide rights of access and inspection address whether a party, usually the buyer, can access facilities to ensure they are operated and maintained properly or as required specifically in the contract. These provisions are critical. In California, for example, utilities have accessed data, files, and facilities to determine whether they are performing as required by law and as specified in the contracts. California independent power providers have lost certain rights as a result. These terms are frequently included in the safe and reliable operations clause or in the metering or interconnection provisions of the contract.

Fuel supply provisions in certain contracts specify how the seller must arrange for fuel supplies. Some such clauses prevent the seller from entering into fuel contracts without review and approval by the buyer. The buyer's approval may not be unreasonably withheld. Some PPAs go into great detail, giving a great deal of discretion to the buyer to control fuel acquisition. This provision may require the seller to have written contracts for fuel supply (including backup fuel) that are satisfactory to the buyer or the contract may be terminated. The contract also may require that alterations to fuel supply and transportation contracts be approved by the buyer's written consent. The PPA may allow the buyer to approve the contracts based on certain factors, including term, price, quantity or quality, availability, transmission and curtailment, and assumability or assignment of rights by buyer.

Remedies for breach

Generally, when a contract is breached, the injured party may cancel the contract and recover (1) the expectation or the value of any performance by that party or (2) specified damages for the breach.[28] The right of the breaching party to cure the contractual default it committed is expressly added in a few contracts, and, where it is added, it gives the defaulting party a period of time to cure deficiencies before breach. If breach is asserted, it must be determined whether the breach is material or partial.[29] At common law, only material breach allows the innocent party the option to cancel. Some of the breach remedies in PPAs include an option to purchase generating facilities, the buyer's right to possess, control, and operate the seller's facilities, the suspension of buyer's obligations, and the forfeiture of deposit or security.

Liquidated damages

Many power sale contracts have liquidated-damages clauses, whereby the parties agree at the time of entering into the contract to specify, in advance, the amount of damages that will be payable if either party breaches the contract.[30] The alternative is to determine the damages that the innocent party incurs when and after the breaching party defaults on its contractual obligations, which is the normal situation. Although there are reasons of certainty to liquidate damages, liquidated damages may be included as a coercive economic penalty to attempt to dissuade a party from certain conduct during the PPA term or from breaching. Typically the stronger party inserts a liquidated-damages clause into the PPA.

However, not all liquidated-damages clauses are enforceable. Only clauses that are fair and reasonable and satisfy legal standards survive. Those that are so unfair or disproportionate as to be unconscionable are not enforceable, even when they are included in the PPA.[31] The legal standard is that a liquidated-damages clause must be a fair and reasonable estimation of damages, not a penalty. This is a factual question that is decided by the court.[32]

In deciding whether a particular clause is enforceable, common law focuses on whether the clause was fair and reasonable at the time the contract was formed.[33] In contrast, the Uniform Commercial Code (UCC) focuses on both the time of contracting and the time of breach; it considers whether the clause was fair and reasonable in light of the damages anticipated at the time of formation and the damages *actually* caused by the breach.[34] This allows a type of hindsight and second guessing. If the liquidated sum appears disproportionate at the time of the breach, courts applying the UCC may strike the liquidated-damages clause and refuse to enforce it (this is discussed in more detail in the next chapter). Therefore, if one employs too aggressive a liquidated amount, or if the amount is not roughly proportional to the actual damages incurred, the clause will not be enforced.

Although many PPAs liquidate damages, this is not necessarily as effective for PPAs as it is for other types of contracts. In a typical contract, the parties may have difficulty determining at the time of contracting what the damages for later breach would be, given the long-term nature of the contract. In some situations, the damages are more easily ascertainable at the time of breach, when, for example, the nonbreaching party has to "cover" or procure power from another source, and thus has an exact amount of damages. At the time of breach, usually there is a course of performance of the contract that demonstrates how the project has performed, how much power it has generated, and how much revenues it has produced.

A PPA is a type of contract in which it is possible to determine the amount of damages at the time of breach; therefore, a liquidated-damages provision in a PPA may be stricken and not enforced. By contrast, a rigid liquidated damages sum typically cannot reflect damages that may vary significantly over the length of a long-term contract. In deciding whether to include a liquidated-damages clause, it is important to consider these issues against the certainty provided by liquidating the amount of damages at the time of contracting.

Liability

Indemnification

This provision of the PPA typically addresses the reimbursement liability of one party resulting from acts of the other. Most PPAs provide for both parties to indemnify, defend, and hold harmless the other party for one's acts or failures to act. Typically, the provision covers bodily injury, death, and property damage caused by either party. The indemnification responsibility is mutual. A typical provision might state, "Each party shall indemnify, save harmless and defend the other, its directors, trustees, agents, officers, and employees, against all claims, demands, judgments and associated costs and expenses, related to property damage, bodily injuries or death suffered by third parties resulting from any act or failure to act by such party related to this Agreement." Some provisions are more broad and cover damages resulting from other losses and expenses. These may include direct, indirect or consequential loss, damage, claim, cost, charge, or expense, including attorney's fees and other costs of litigation.

The most general PPA provisions simply designate each party as an individual unrelated party, without specifically requiring indemnification. A typical provision might state, "Nothing in this Agreement shall be construed to...affect the status of (the buyer) as an independent public utility corporation, or Seller as an independent individual or entity." This leaves the parties to ask for indemnification at common law when it is not specifically provided by the contract. The most specific PPAs authorize many types of damages, including attorney costs, to be included in this action for indemnification. Some indemnification provisions only ask for defense of claims on request. Some provisions include a proportionality term, which specifies that expenses will be shared proportionally if negligence is shared.

Liquidated-damages provisions typically cover specific damages owed to the buyer in the event the seller breaches. One type of provision is triggered if the buyer terminates; some account for years of project operation, whereas others let the buyer keep adjustments, offsets, and refunds as liquidated damages.

Lenders' Rights

A third party to a contract only acquires the right to enforce the contract if the court finds that the principal parties to the contract intended to create legally enforceable rights in the third party. The critical test is whether this creation of third-party rights was intended by the contracting parties at the time the PPA was formed.[35] An intended third-party beneficiary to these contracts can enforce its intended rights under the contract.[36] A key group of designated or intended third-party beneficiaries of the PPA are the lenders that finance the project. When lenders are anticipated, PPAs provide for the power seller's lenders to be empowered to cure a seller's breach. A typical provision might provide, "The Project Lender may cure the default on behalf of Seller and continue the Agreement with (Buyer) if the Project Lender assumes all of Seller's rights and obligations arising under this agreement."

Some contracts require that notification of breach by the seller also be given to the seller's financiers. Other parties that, either expressly or by implication, are not intended third-party beneficiaries of the PPA have no rights thereunder. Liens to secure amounts owed may be perfected against an independent project without any express third-party PPA provisions by contractors, subcontractors, suppliers, and material men who provide services or supplies to the project and are not paid.

Notes

1. To a lesser degree, some contracts are based on a UK model of financing. If the Power Purchase Agreement (PPA) is not in an official English-language version, this impedes international and, in some cases, local project finance.

2. Corbin, Arthur Linton. 1952. *Corbin on contracts.* Vol. 1. St. Paul, MN: West Publishing. Corbin states that "such mutuality is necessary is one of the most commonly reported statements known in the law of contract. Both parties to a contract must be bound or neither is bound. The contract is void for lack of mutuality. Thus the law is stated."

3. Ibid., 95. Corbin states, "Vagueness of expression, indefiniteness and uncertainty as to any of the essential terms of an agreement may prevent the creation of an enforceable contract."

4. Information required may include providing status within 15 minutes of a request, a daily plan for the next day, weekly updates to the annual outage and maintenance plans, explanation of cause of outage of circuit breakers, and annual monthly generation estimates with prompt updates.

5. National Independent Energy Producers (NIEP). 1992. *Negotiating risk: Efficiency and risk sharing in electric power markets.* Washington, DC: NIEP.

6. Ferrey, Steven. 2005. *The law of independent power,* 22nd ed. St. Paul, MN: West Publishing.

7. NIEP, *Negotiating risk*, 11.

8. Ibid., 11.

9. *American jurisprudence* 2nd ed. at 117A: sec. 380 supports this by stating that "there is an implied undertaking in every contract on the part of each party that they will not intentionally and purposely do anything…which will have the effect of destroying or injuring the right of the other party to receive the fruits of the contract."

10. Corbin, *Corbin on contracts*, sec. 973. Corbin states, "In order to constitute an anticipatory breach of contract, there must be a definite and unequivocal manifestation of intention on the part of the repudiator that he will not render the promised performance when the time fixed for it in the contract arrives."

11. Corbin, *Corbin on contracts*, sec. 752; *American jurisprudence* sec. 526.

12. *American jurisprudence* 1: sec. 513.

13. See Chapter 5 of this volume.

14. NIEP, *Negotiating risk*, 8.

15. *American jurisprudence* sec. 655.

16. Ibid. sec. 657 states that "no consideration moving between the parties is necessary to support a waiver of a contractual provision."

17. Ibid. sec. 527. "The rule followed by the courts generally, with some authority to the contrary, is that a written contract not required by the law to be in writing may be modified by a subsequent oral agreement even though it provides that it can be modified only by a written agreement."

18. Corbin, *Corbin on contracts*, sec. 866.

19. Ibid., sec. 866. Corbin also states that "the doctrine that a provision in a contract forbidding the assignment of rights is invalid because it involves an improper restraint on the alienation of property."

20. Corbin, *Corbin on contracts*, sec. 923.

21. Ibid., sec. 1333; *American jurisprudence* sec. 673 and sec. 676.

22. A typical provision might read, "Force Majeure is limited to unforeseeable causes beyond the reasonable control of and without the fault or negligence of the party claiming Force Majeure, and specifically excludes strikes, walkouts, lockouts or other labor disputes in which Seller's employees participate as well as nonavailability of fuel to operate Seller's Facility. If either party is rendered wholly or partly unable to perform its obligation under this Agreement because of Force Majeure, that party shall be excused from whatever performance is affected by the Force Majeure to the extent so affected."

23. NIEP, *Negotiating risk,* 13.

24. Ibid., 14.

25. Ibid., 13.

26. Ibid., 15.

27. *American jurisprudence* sec. 626. "Later authorities require merely substantial, rather than a strict or literal, performance of a contract, subject to a right of recoupment in respect of the variation of the actual from the stipulated performance."

28. *American jurisprudence* sec. 724.

29. Ibid., sec. 726. Determining whether a breach is "total or partial is key. A material breach, as where the breach goes to the whole consideration of the contract, gives to the injured party the right to rescind the contract or to treat it as a breach of the entire contract… and to maintain an action for damages for a total breach…Even a partial breach of a contract ordinarily gives a right to damages, although the contract remains in effect."

30. *American jurisprudence* 2nd ed. 22: sec. 683.

31. *Priebe and Sons v. United States,* 332 U.S. 407 (1947).

32. *Knotton v. Cofield,* 160 S.E. 2d 29 (N.C. 1968).

33. *Truck Rent-a-Center v. Puritan Farms, Inc.*, 361 N.E. 2d 1015, 1016 (N.Y. 1977).

34. *Equitable Lumber Corp. v. IPA Land Dev. Corp.*, 344 N.E. 2d 391, 395 (N.Y. 1976). For more on differences between the common law and the UCC on legal decision rules, see chapter 12.

35. *American jurisprudence* sec. 440.

36. Ibid., sec. 435 states that "It is well established that a third person may, in his own right and name, enforce a promise made for his benefit even though he is a stranger both to the contract and to the consideration."

12 The Law and Principles Governing International Power

Securing Power Contract Enforcement

The role of the power purchase agreement in capital flows

The key contract provisions examined in chapter 11 mean nothing if they cannot be enforced promptly and efficiently in a neutral forum. International interest in power sale deals will only advance, and international capital will only flow to such deals if there is confidence that power purchase agreements (PPAs) embodying those deals are fair and enforceable. Because international lenders and developers view many developing countries as having relatively immature judicial systems, as not having a long-established independent regulator that can enforce PPAs independent of utility and government agencies, or not having established a system of competitive retail sale of power, there is hesitation among SPP developers and financiers.

Power generation technologies are hard assets that are bolted to the ground. Once installed, power generation plants cannot easily be picked up and taken away if a dispute or impasse occurs. This is especially true for fossil fuel-fired facilities and hydroelectric facilities, which require a significant civil works component (cement, dams, weirs, impoundments, etc.) that cannot be moved once it is installed.

During the Asian financial crisis in the late 1990s, when power buyers did not want to continue purchasing power at contractually set prices or at contractually set terms, there were significant problems with the enforcement of some existing power sale agreements for independent power producers (IPPs) and small power producers (SPPs) in Indonesia, India, and elsewhere. An absolute "take or pay" purchase obligation, a power purchase price that is linked to changes in the U.S. dollar exchange rate against the local currency, and the rights of the SPP to sell power directly to retail buyers are contractual provisions that must be backed up by an independent regulator in the country or enforced by a court of law to be effective. Therefore, the SPP developer must have confidence in the legal enforceability of the PPA. The PPA is the legal living "skin" of the power project: It is what makes the project a viable entity.

The risks posed by single-buyer electric sectors and choice of forum

A significant barrier to SPP investment stems from the fact that almost all developing nations have a single utility that is government owned as the only authorized power purchaser. This dramatically narrows the option for the sale of power output to a single entity, which may be compelled by the government to charge less for retail power tariffs than its cost of production (and therefore running at a loss), may not be independently creditworthy, and may be subject to annual government appropriations for a significant part of its revenue to pay for purchased power. That utility, as a government-affiliated entity, is perceived as not being on a level playing field with an SPP should there be a contract dispute that ends up in the local courts. Therefore, SPP developers are keenly aware of whether the power system is structured to allow them other power sale options and whether the PPA will be enforceable in a neutral forum under the acceptable rules of law.

First, some lenders attempt to obtain credit or equity guarantees from multilateral agencies or host countries. As discussed in chapter 10, parts of the World Bank can provide commercial and political risk guarantees and credit support. In addition, the countries from which SPP or IPP power generation equipment is sourced can provide Export-Import Bank or other guarantees for loans so as to promote the purchase of equipment from the host country. These provide a second line of credit support.

However, to invoke payment under this credit support, loan documents typically require that the SPP or IPP owner pursue to finality judicial proceedings against the defaulting party in a neutral international forum.

Second, the system may allow some alternatives for power sale. The previous chapters examining the five specific Asian countries highlighted whether direct retail sales of power output or net metering (energy banking) were permitted. Either of these elements provides more confidence to an SPP developer, investor, or lender.

But systems can and do change. Where such a system of direct retail sales of power is allowed, it can be revoked, as in California. Where net metering is allowed, it can be revoked, as in some of the states of India. Ultimately, the investment is only as good as its *enforceable* guarantees, which are embodied in the PPA.

Third, SPP and IPP developers, as well as their lenders, typically try to provide in the PPA for arbitration or court adjudication outside the developing nation in a neutral forum, where there is a longer history of such proceedings and prompt resolution. Delay in resolution works against the SPP owner, which typically owns immovable power generation assets and is constrained to selling power only to the national utility in the host developing nation. For Asian SPPs, developers and lenders often favor Singapore, a center of Asian capital markets, for such proceedings. The Hague, Brussels, or Geneva, all recognized international venues with a history of acting as hosts for multinational proceedings, are also common choices, but obviously they not in the immediate vicinity of many developing nations. National utilities usually resist such out-of-country forums: It is a matter of national pride. In addition, government-owned utilities often are quite confident in the favorable results they can achieve in a local court.

To adjudicate U.S. contract disputes, it is typical that the forum will be located in a place where either party has its place of business or where the contract was made. However, if the parties designate the choice of forum in the PPA, this usually is respected by the courts.[1] This underscores how important it is for SPPs to have a neutral, standardized PPA that is fair and neutral in all of its operative provisions.

Fourth, the parties to the PPA may attempt by contract to designate what substantive law applies to arbitrate or adjudicate a later dispute involving the PPA. This is different from designating *where* that adjudication will take place. Nations have very different legal systems and substantive laws. Just the five developing Asian nations profiled in the earlier chapters exemplify an Islamic legal system (Indonesia), a communist/socialist system (Vietnam), a system based in part on English legal traditions and in part on Hindi law (India), and a system that is distinct from all of the above (Thailand). There are substantive differences.

The PPA may designate which system of laws will apply, irrespective of where the venue is established to conduct the dispute resolution. Some local courts in developing nations, as a matter of legal practice, will only apply local law, rather than international law or the law designated in the PPA, to adjudicate disputes. Again, this underscores the importance of utilizing a neutral, standardized PPA.

Whose law applies?

International lenders require that their loans and the PPA establish recourse to a mature system of substantive law that is perceived as offering neutral and expeditious resolution. As most international lenders are based in the United States or Europe, that system of required law is either (1) U.S. law, (2) UNIDROIT (International Institute for the Unification of Private Law), a code of international contract laws, or, to a lesser extent, (3) English law. There is a second set of international contract law principles, known as the Convention on the International Sale of Goods (CISG). However, it is specifically stated in the CISG that these principles are not meant to apply to the sale of electricity.[2] Therefore, UNIDROIT is the only potentially applicable international law.

When U.S. law applies to adjudicate PPA disputes, there are two substantive systems of law that apply to contract disputes: the common law and the Uniform Commercial Code (UCC). Article 2 of the UCC is a model statute adopted in every U.S. state[3] that is designed to streamline the legal treatment of sales of "goods." Whether contracts for the sale of electricity are governed by the UCC or by common law depends on whether electricity is determined to be a good, as defined by

Section 2-105 of the UCC. The definition states, in part, that goods are (1) all things (including specially manufactured goods) that are movable at the time of identification to the contract for sale and (2) must be both existing and identified before any interest in them can pass.[4]

Different states have wrestled with this definition and come to different conclusions. Some jurisdictions, such as New York, treat electricity strictly as a service, whereas other jurisdictions have drawn distinctions between "raw" energy and "metered" energy, treating the former as a service and the latter as a good. Still other states, such as California, have reached inconsistent and varying determinations at different times within the state as to whether electricity is a good or a service.

Rules in evolving electricity markets will be made by contract and interpreted by courts; which substantive law applies will depend on whether electricity is characterized as a good or service.[5] It is important to analyze the strategic differences between the common law of contracts and the statutory provisions of the UCC on electricity transactions and contrast these with UNIDROIT principles.

There are several important differences among common law, the UCC, and UNIDROIT. Table 12–1 illustrates those key differences. The following sections analyze the key differences in interpreting PPAs under common law, the UCC, and the international UNIDROIT principles.

Table 12–1: Key Difference in Law for Electricity PPAs

Issue	Common Law	UCC	UNIDROIT
1. Must acceptance of a contract offer exactly "mirror" the terms of the offer?	Yes	No	Often not
2. Can additional terms to the deal be added by the acceptance, even if not contained in the offer, in wholesale electric transactions?	No	Yes	Yes
3. Must enforceable contracts for more than $500 either be in writing or evidenced by a writing?	No	Yes	No
4. Can an existing contract be modified without new consideration?	No	Yes	Yes
5. Are prior oral statements includable as part of a written contract?	If consistent	If consistent, as well as conduct and trade usage	If consistent
6. Can a contract be orally modified even where that contract prevents such modification?	Yes	No	No
7. Will indefinite gaps in a contract be filled in and the contract enforced?	Often not	Usually	Yes
8. Must a demand for assurances of performance be in writing? Is response always required in less than 30 days?	No	Yes	No
9. Can a firm offer in writing not supported by consideration be revoked?	Yes	No	No
10. Is substantial performance of obligations, rather than perfect performance, allowed?	Yes	No	Possibly
11. Are trade practices and past conduct relevant in interpreting the deal?	Often not	Always	Yes
12. Is the market value of an item measured at the time of breach rather than at the time the party was to perform?	Yes	No	Yes
13. Will implied warranties of merchantability and fitness be read into the contract?	No	Yes	No
14. If the warranty remedy fails, will courts throw out quality disclaimers?	No	Yes	No

Form, Formation, and Modification of Power Sale Contracts

Formation of the power purchase agreement

The law surrounding power purchase contracts that contain indefinite or ambiguous terms differs among common law, the UCC, and UNIDROIT. Indefiniteness can occur when parties leave material terms of the deal open or not fully specified in the PPA.[6] Indefiniteness is grounds for negating the enforceability of an otherwise properly formed PPA contract.[7]

Common law. Under the *Restatement (Second) of Contracts*, common law requires mutual assent for contract formation. The manifestation of mutual assent to a deal ordinarily takes the form of an offer or proposal by one party, followed by an acceptance by the other party; it may be made even though neither an offer nor an acceptance can be identified and even though the moment of formation cannot be determined.[8] The fact that one or more terms of a proposed bargain is uncertain may show that a manifestation of intention is not desired to be understood as an offer or as an acceptance of the deal.[9]

Indefiniteness typically arises in three situations: (1) when the parties have purported to agree on a material term but have left it indefinite; (2) when the parties are silent as to a material term; and (3) when the parties have agreed to agree on a material term. In the first situation, at common law there is no room for judicially inserted "gap-fillers" (a term that a court supplies); therefore, the agreement is void.[10] In second situation, there is the possibility that the term may be implied from the surrounding circumstances or supplied by a court using a gap-filler. In third situation, the traditional common law rule is that an agreement to agree does not result in a binding PPA. Some courts do enforce agreements to agree, or at least require good faith efforts of the parties to agree subsequently.

UCC. The formalistic common law approach differs greatly from the UCC standard. As to indefiniteness, the UCC provides that a contract for the sale of goods may be made in any manner that is sufficient to show agreement, including conduct by both parties that recognizes the existence of such a contract.[11] Even though one or more terms are left open, a contract for the sale of goods does *not* fail for indefiniteness if the parties intended to make a contract and if there is any reasonably certain basis for the tribunal giving an appropriate remedy. In other words, under the UCC, the court can cure the indefiniteness on any reasonable basis and enforce the contract. Under the UCC, the court fills the gaps with a reasonable price at the time for delivery if (1) nothing is said as to price, (2) the price is left to be agreed by the parties and they fail to agree, or (3) the price is to be fixed in terms of some agreed market standard set by a third person and is not so set.[12]

An agreement for sale that is otherwise sufficiently definite to be a contract is not made invalid by the fact that it leaves particulars of performance to be specified by one of the parties.[13] This allows contracts in which price, quantity, or another material term in an electric sale are subject to contingencies or to one party's specifications. However, any such specifications must be made in good faith and within the limits of commercial reasonableness.[14] The use of any quantity estimate in an agreement becomes the midpoint of a reasonable quantity range for the sale of electricity. The parties as they so intend can conclude a contract for sale even though the price is not settled. In such a case, the price is a reasonable price at the time for delivery if (1) nothing is said as to price, (2) the price is left to be agreed by the parties and they fail to agree, or (3) the price is to be fixed in terms of some agreed market or other standard as set or recorded by a third person or agency and it is not so set or recorded, (4) a price to be fixed by the seller or by the buyer means a price for him to fix in good faith, (5) a price left to be fixed otherwise than by agreement of the parties fails to be fixed through fault of one party, and the other may, at his option, treat the contract as cancelled or himself fix a reasonable price, (6) the parties intend not to be bound unless the price be fixed or agreed, and it is not fixed or agreed, there is no contract.[15] Many electricity contracts are requirements contracts in which the quantity is determined by the reasonable needs of the buyer.

UNIDROIT. The UNIDROIT principles take the most liberal position, like the UCC, in allowing the tribunal to fill in gaps in the PPA and allowing the PPA to be enforced: "Where the parties to a contract have not agreed with respect to a term, which is important for a determination of their rights and duties, a term which is appropriate in the circumstances shall be supplied. In determining what is an appropriate term regard shall be had, among other factors, to (a) the intention of the parties, (b) the nature and purpose of the contract, (c) good faith and fair dealing; and (d) reasonableness."[16]

Option contracts

Option contracts and a variety of hedge "put" and "call" options are commonly used in the electricity markets. Option contracts are an offer for a deal that is irrevocable for a period of time.[17] Irrevocable offers may be terminated by a lapse of time, the death or destruction of a person or thing that is essential for performance, or a supervening legal prohibition.[18]

Common law. The UCC and common law differ in their treatment of option contracts. Under the majority common law position, option contracts need new, mutual, and distinct consideration to be enforceable.[19] Therefore, without separate consideration (or estoppel) at common law, in many states, an option is gratuitous and an offer is not irrevocable and may be withdrawn. Electricity options require consideration to be enforceable if electricity is deemed a service.

UCC. The UCC, however, empowers a party to create an irrevocable offer for the sale of goods without consideration.[20] Under the UCC, the requisites are that (1) the offerer must be a merchant, (2) the offer must be in a signed writing, (3) if the language of irrevocability is on a form supplied by the offeree, the offerer must sign twice—once to make the offer and once more to sign the clause providing for irrevocability, (4) the writing must contain language of irrevocability, and (5) the period of irrevocability may not exceed three months.[21]

UNIDROIT. The UNDIROIT principles make options irrevocable for the time stated, even without purchasing that option for consideration: "[A]n offer cannot be revoked if it indicates, whether by stating a fixed time for acceptance or otherwise, that it is irrevocable."[22] Thus, it is similar to the UCC provision, but more liberal and without the strict UCC requirements.

Acceptance of offers to sell electricity

What constitutes "acceptance" of a deal for power also varies between common law and the UCC. Contracts are a procedural two-step choreography. Upon the second step, a valid acceptance of an offer, a contract is formed.

Common law. At common law, an offer can be accepted only in the manner that the offer is made: If the offer solicits a return promise, it can be accepted only by making that reciprocal promise. Acceptance by the rendering of the requested performance can occur only if the offer invites such an acceptance (that is, unilaterally).[23] At common law, it matters as to whether a deal is properly formed whether one accepts that deal by return promise or by conduct.[24]

UCC. The UCC differs from the common law rule by eliminating the distinction between promise and conduct: An offer to enter into a contract is construed as inviting acceptance in any manner and by any medium that is reasonable under the circumstances.[25] Either words or conduct may be appropriate, under the circumstances, to manifest an acceptance if electricity is deemed to be a good and is governed by the UCC.

Additional terms added to the power purchase agreement

Common law, the UCC, and UNIDROIT differ significantly as to whether additional or varying terms may be inserted into the PPA by an accepting party to the deal (offeree) as part of its acceptance.

Common law. The majority common law position requires an acceptance to be substantially a mirror image of the offer.[26] The manner and mode of acceptance also must be reasonable and as reliable as the mode of communication of the offer. To do otherwise is not an acceptance, but an implied rejection and a counteroffer.[27] An offeree's power of acceptance is terminated by the intentional or unintentional making of a counteroffer,[28] unless there is an open option (discussed previously) benefiting the offeree. Thus, at common law, the technical conformance of an acceptance as a mirror image of the offer, substantively and procedurally, is critical to forming a PPA.[29] Additional terms cannot be inserted into an agreement by the offeree as part of the acceptance.

UCC. The UCC has a very different and more flexible approach. It provides that a definite and seasonable expression of acceptance[30] or a written confirmation, sent within a reasonable (or the specified) time, operates as an acceptance, even though it states terms that are additional to or different from those offered or agreed upon.[31] If one party to the deal is not a merchant, the additional terms are mere proposals for additional terms, which are not incorporated unless subsequently assented to by the offerer. Between two or more electricity merchants,[32] such additional terms become part of the contract automatically unless it is materially altered, or unless the offerer initially limits acceptance to the express terms after seeing such terms or "seasonably"[33] objects.[34] In addition, if conduct by both parties recognizes the existence of a contract, such conduct is sufficient under the UCC for a transaction in goods to establish a contract for sale, although the writings of the parties do not otherwise establish a contract.[35] Therefore, it is much easier to form a PPA under the UCC, notwithstanding nonconforming terms in the acceptance.

UNIDROIT. Under UNIDROIT, the acceptance of a contract offer needs to mirror the terms of the offer: "A reply to an offer which purports to be an acceptance but contains additions, limitations, or other modifications is a rejection of the offer and constitutes a counter-offer."[36] This is similar to the common law American rule. However, a "reply to an offer which purports to be an acceptance but contains additional or different terms which do not materially alter the terms of the offer constitutes an acceptance, unless the offerer, without undue delay, objects to the discrepancy. If the offerer does not object, the terms of the contract are the terms of the offer with the modifications contained in the acceptance."[37]

The UNIDROIT principles thus create a hybrid between the common law and UCC rules. Under UNIDROIT, (1) an acceptance must mirror the offer, but (2) it can add nonmaterial changes or additions, unless (3) the other party notices and objects to such nonmirrored acceptance, in which case the nonmirrored acceptance is not a valid acceptance. Under UNIDROIT, there is no distinction between merchants and consumer transactions, as in the UCC. It takes a middle position and leaves the court or tribunal to decide what is a material addition or change and what is minor.

Modifications to the PPA

Common law. Rules on the modification of an existing PPA are yet another area where the bodies of law conflict. For a modification, the common law majority opinion holds that new mutual consideration is required for each modification.[38] Therefore, a change in an electricity service contract is not enforceable at common law unless it is supported by new mutual consideration.

UCC. Under the UCC, however, new agreements modifying provisions of an existing PPA need no new consideration to be binding.[39] Such modification may be oral, except in two situations: (1) when the agreement as modified is required to be evidenced by writing under the statute of frauds (discussed next), and (2) when a writing is required under Section 2-209(2).[40]

UNIDROIT. The UNIDROIT principles comport with the UCC position: "A contract is...modified...by the mere agreement of the parties, without any further requirement."[41] Moreover, under the UCC, a signed PPA stating that it cannot be modified or rescinded except by a signed writing, contrary to the rule under the common law, will be honored.[42] Under UNIDROIT principles, a similar rule is followed: "A contract...which contains a clause requiring any modification...to be in writing...may not be otherwise modified. However, a party may be precluded by its conduct from asserting such a clause to the extent that the other party has acted in reliance on that conduct."[43] If electricity is considered a service under common law, such a limitation on oral modification is not enforceable. Thus, if the parties to a PPA modify a term orally or in a phone conversation, that modification is enforceable only under common law.

Must the PPA be evidenced in writing? The statute of frauds

Common law. One of the key distinctions among the various rules of law is the requirements of the statute of frauds. At common law, electricity contracts that are not to be performed fully within one year, PPAs that involve a suretyship or guarantee of payment, and, in some states, PPAs that set up an agent or broker relationship fall within the statute of frauds and require written evidence of the deal to be enforceable in a court.[44] Simply stated, the statute of frauds is a long-established principle that certain deals require written evidence of the deal for the contract to be enforceable. At common law, the party seeking to enforce the PPA must be able to produce a writing (or multiple writings with common reference), containing all material terms, that evidences the signature of the party against which the contract is to be enforced (in other words, the party that is seeking not to be bound by the deal).[45]

UCC. In contrast, Section 2-201 of the UCC states that for a contract for the sale of goods over $500 to be enforceable, there must be some writing(s) sufficient to indicate that a contract for sale has been made.[46] All that is required is that the writing(s) contain the signature of the party to be bound and the quantity of the good and that it be a contract for the sale of goods. The price, time, and place of payment or delivery, the general quality of the goods, or any particular warranties may all be omitted from the writing and filled in by the court. Moreover, under the UCC, a writing is not insufficient because it incorrectly states a term that has been agreed upon; if the quantity of electricity to be sold is understated, however, recovery is limited to the amount stated.[47]

The UCC also has a "two merchants" exception:[48] When it applies, it makes a writing sent by one merchant to another sufficient to defeat a statute of frauds defense asserted by either party to the PPA. Therefore, that singularly signed writing binds both parties to the PPA unless the recipient of the writing seasonably objects within 10 days of receipt of the writing. Common law has no such reciprocity provision: The PPA can be binding against one party, whereas the other has a statute of frauds defense. The UCC also takes out of the statute of frauds any contracts for specially manufactured goods and contracts fully performed or admitted to.[49] Therefore, the requirement of a writing to authenticate a power deal is more easily satisfied if electricity is deemed a good and the UCC applies.

General principles. What qualifies as a signature and satisfies the requirement of the statute of frauds has evolved throughout the years. Today, a signature "is any mark, written, stamped or engraved that is placed by a party anywhere on the writing with intent to assent to and adopt (authenticate) the writing as the party's own."[50] For example, initials and rubber stamps are sufficient. Letterhead may be a signature. When the terms necessary to satisfy the statute are present in two or more documents and only one is signed by the party to be bound to the PPA, if the unsigned document is physically attached to the signed document at the time it is signed, or if one of the documents expressly refers to the other by its terms, the statute is satisfied and the PPA is enforceable.[51]

UNIDROIT. Under UNIDROIT, there is no statute of frauds issue at all. There is no requirement that certain contracts be evidenced by a writing to be enforceable: "Nothing in these Principles requires a contract to be concluded in or evidenced by writing. It may be proved by any means, including witnesses."[52] There are different statute of frauds standards under common law (if PPAs, brokers, or guarantors are involved) and the UCC (if the value is over $500), but no statute of frauds requirements under UNIDROIT. Therefore, if the PPA is not written or evidenced by a writing, there may be technical legal reasons why it is not enforceable even if it is otherwise in order under common law or the UCC.

Interpretation of the Deal

Parol evidence rule

In many deals it is typical for the salesperson to make significant claims to close the deal, but, when the parties follow up by signing a written PPA, it may not contain those aspects promised by the salesperson. In fact, it may even disavow or disclaim some of those oral promises. When a dispute arises between the parties and these oral promises are critical, do they count as part of the deal if the written PPA is silent or contradictory about this term?

This is a critical issue in many PPAs. It is resolved by a very complex concept known as the parol evidence rule. The rule differs at common law and under the UCC. The parol evidence rule excludes evidence of certain terms agreed upon before or contemporaneously with an integrated written PPA. [53] The more complete and formal the instrument, the more likely that it is intended to be an integration, in written form, of the entire agreement, and therefore prior oral evidence is not allowed.

Once it is determined that a writing is the final embodiment of the deal, the second question is whether the writing is complete so that it is a total integration, exclusive of parol evidence. Different precedents determine this in distinct ways. Williston, Corbin, and the UCC have articulated distinct views. Professor Williston's view has three rules: If the writing expressly states that it is a final expression of all the terms agreed upon and that it is a complete and exclusive statement of these terms (referred to as a "merger clause"), this declaration conclusively establishes that the integration is a total integration unless the document is obviously incomplete or the merger clause itself was included as a result of fraud or mistake. His second rule states that, in the absence of a merger clause, the determination of whether the writing is a total integration is made by looking to the writing. Consistent additional terms may be proved if the writing is obviously incomplete on its face; where the writing appears to be a complete instrument expressing the rights and obligations of both parties, the writing is deemed a total integration. A varying approach in determining the completeness of a writing is Professor Corbin's view. Professor Corbin is determined to search for the actual subjective intent of the parties on this issue of intended total integration rather than some objective or presumed intent. He states that all relevant evidence, including parol evidence, should be taken into account in making this threshold determination. Thus, he would admit evidence of prior negotiations. Furthermore, he states that a merger clause is only one of the factors to be considered in determining whether there is a total integration.[54]

Common law. Under the common law rule, even if there is a determination that the parties intended the final form of the power deal to be embodied in the written PPA, consistent additional oral or prior terms are still admissible if they are consistent with the written PPA and if the offered terms might naturally, ordinarily, or normally be omitted from the PPA that does exist.[55]

UCC. The UCC embodies another view. The UCC approach creates a presumption that a written PPA is only a partial integration of the deal.[56] In such situations, the following oral or prior terms may be produced as evidence at trial:

- Terms that are consistent with the written PPA[57]

- Terms that "certainly" would not have been included by the parties in the written PPA

- Course of dealing,[58] course of performance,[59] and trade use[60]

If the parties so designate in their written PPA, it can be deemed a complete and exclusive embodiment (full integration) of the final deal. This essentially provides that not only is the written PPA the embodiment of the final terms of the deal, but also it expressly negates any oral or prior agreements, which may not be offered as evidence to a court or arbitrator resolving a PPA dispute between the parties. If the written PPA is determined to be fully integrated—complete and exclusive—most courts will only admit as evidence, other than the PPA itself, course of dealing, course of performance, and trade use.

Nonetheless, more parol (oral or prior) evidence of the nature of the power deal can be admitted if electricity is deemed a good under the UCC than if it is deemed a service under the rule of common law. The UCC will always allow the parties' conduct, their past dealings, and industry norms to be used by a party to gain a favorable interpretation of a contract. This allows contextual evidence that enables a judge to interpret the deal even when the parties thought that only the written contract set forth their legal relationship.

Under the UCC, regardless of what is written in the PPA, conduct matters. How one conducts oneself in the early stages of performance of an electricity deal may be used as evidence. The UCC creates an interpretive hierarchy. The UCC statutorily provides that express terms (in the PPA) shall "control"[61] over the course of performance, and that course of performance shall "control" over the course of dealing or usage of trade.[62] Thus stated, terms dominate over past conduct under the agreement,[63] which, in turn, dominates over conduct under past deals, which, in turn, dominates over industry and trade norms—but all can be employed to interpret obligations if electricity is deemed to be a good under the UCC. A course of dealing and course of performance can be established by the testimony of the parties, but expert witnesses usually prove trade usage.[64] Under the UCC, the conduct of the parties and, within the industry, trade practices generally, informs the parties' performance obligations with electricity goods. With electricity services, conduct and industry practices are less dispositive.

UNIDROIT. Trade practices and past conduct are very relevant in interpreting the PPA with UNIDROIT principles: "In applying Articles 4.1 ['Intention of the parties'] and 4.2 ['Interpretation of statements and other conduct'], regard shall be had to all the circumstances, including (a) preliminary negotiations between the parties, (b) practices which the parties have established between themselves, (c) the conduct of the parties subsequent to the conclusion of the contract, (d) the nature and purpose of the contract, (e) the meaning commonly given to terms and expressions in the trade concerned, (f) usages."[65] In this sense, it is similar to the UCC provisions.

Under the UNIDROIT principles, it is possible to prevent prior oral or other evidence from being used to contradict or even supplement the terms of the written PPA. However, that prior evidence may be used to interpret the meaning of the words and the phrases in the written PPA, as long as they are not contradicted: "A contract in writing which contains a clause indicating that the writing completely embodies the terms on which the parties have agreed cannot be contradicted or supplemented by evidence of prior statements. However, such statements may be used to interpret the writing."[66] Thus, UNIDROIT is similar to the common law position on parol evidence.

Power warranties

UCC. One of the most significant differences between the common law of contracts and the UCC is the warranties that are automatically imposed on a deal if it involves the sale of goods under the UCC. These warranties might apply to the quality, quantity, and uniformity of delivering the electricity to be sold. One automatically implied warranty is a warranty that the goods shall be merchantable if the seller is a merchant with respect to goods of that kind, unless the warranty is effectively disclaimed as part of the power deal. Goods to be merchantable must "pass under the contract description" and be "fit for the ordinary purposes for which such goods are used."[67]

A second implied warranty is the implied warranty of fitness for a particular purpose. It is created when, at the time of contracting, the seller has reason to know any particular purpose for which the goods are required, and the buyer is relying on the seller's skill or judgment to select or furnish suitable goods. This could apply to the need of a buyer for a very reliable power supply. Unless this is conspicuously excluded or modified as part of the deal, an implied warranty that the goods shall be fit for such a purpose is created automatically.[68] To exclude or modify any implied warranty of fitness, the exclusion must be embodied in a writing and conspicuous.[69]

Therefore, when electricity is deemed a good, failure to provide proper voltage or amounts of electricity may violate an implied warranty unless that warranty was properly and conspicuously disclaimed in the PPA. When circumstances cause an exclusive or limited remedy to fail its essential purpose, the two implied warranties are reinstated under the UCC.

UNIDROIT. Under UNIDROIT, no implied warranties of merchantability or fitness are read into the contract. However, a standard of reasonable quality is applied to each party's obligations to perform: "Where the quality of performance is neither fixed by, nor determinable from, the contract a party is bound to render a performance of a quality that is reasonable and not less than average in the circumstances."[70] Therefore, there is a standard of satisfying the industry norm in one's fulfillment of obligations. What constitutes industry norms for electricity at a particular time and place would be established by expert witnesses.

The Obligation to Honor the Power Purchase Agreement

Assurances the power purchase agreement will be performed

When one party does not perform under the power purchase agreement, the other party has the ability to suspend its performance and demand legal assurances of the nonperforming party. The right to seek assurances has different requirements at common law, under the UCC, and pursuant to UNIDROIT.

Common law. Under the *Restatement (Second) of Contracts* at common law, when one party has reasonable grounds to believe that the other party will commit a breach by nonperformance, the first party may demand adequate assurances of due performance. The demand need not be made in writing. That party may treat the failure to provide such assurances within a "reasonable time" as a repudiation of the PPA.[71]

UCC. The UCC also allows a demand for assurances, but when reasonable grounds for insecurity arise with respect to performance, *either party* may demand adequate assurances of due performance in *writing*.[72] First, the aggrieved party is permitted to suspend its own performance and any preparation therefore, with excuse for any resulting necessary delay, until the situation has been clarified.[73] Second, the aggrieved party is given the right to require, in writing, adequate assurances that the other party's performance will be duly forthcoming.[74] Finally, the UCC provides that the aggrieved party may treat the contract as breached if reasonable grounds for insecurity are not cleared up within a "reasonable" time, never exceeding 30 days.[75]

UNIDROIT. The UNIDROIT principles are similar to the less demanding common law provisions: "A party who reasonably believes that there will be a fundamental nonperformance by the other party may demand adequate assurances of due performance and may meanwhile withhold its own performance. Where this assurance is not provided within a reasonable time the party demanding it may terminate the contract."[76]

How exact must performance be?

The standards of performance of the PPA also differ among common law, the UCC, and UNIDROIT. Whether performance needs to be exact or merely substantially in accord with PPA provisions depends on whether a common law transaction in services or a UCC transaction in goods is involved in the electricity sale.

Common law. The majority common law rule holds that substantial, but not exact, performance is sufficient to prevent a material breach of the PPA.[77] Substantial performance short of complete performance is still a breach, but a nonmaterial breach. This means that some monetary damages may be awarded to compensate the nonbreaching party, but the nonbreaching party may not suspend its performance obligations. Only substantial performance is recognized and legally significant under common law.[78]

UCC. The UCC, however, provides that if the goods or their delivery fail to conform to the PPA requirements in any respect, the buyer may reject or accept all or any part of the delivery.[79] This "perfect tender" rule is directly contrary to the common law concept of substantial performance. The seller of power has a right to cure defective performance within a reasonable time,[80] even after the buyer's acceptance of the electricity.[81] To make an effective rejection, the buyer must seasonably notify the seller of the rejection and state all defects that are discoverable by reasonable inspection. For the purposes of the sale of electricity, which travels at almost the speed of light, tender must be exact and perfect if electricity is a good. Substantial performance is satisfactory if electricity is a service.

UNIDROIT. The UNIDROIT principles invoke yet a third rule. The key inquiries are whether the nonperformance goes to the essence of the contract and whether the failure was intentional: "A party may terminate the contract where the failure of the other party to perform an obligation under the contract amounts to a fundamental nonperformance. In determining [this] regard shall be had, in particular, to whether.... (a) the nonperformance substantially deprived the party of what it was entitled to expect under the contract...(b) strict compliance with the obligation which has not been performed is of essence under the contract; (c) the nonperformance is intentional or reckless;...(d) the nonperformance gives the aggrieved party reason to believe that it cannot rely on the other party's future performance; (e) the nonperforming party will suffer disproportionate loss as a result of the preparation or performance if the contract is terminated."[82]

The essence of the contract is license for the tribunal to search for the true purpose of the contract and whether that purpose has been frustrated. At common law, a material breach is often determined by the degree to which the nonperformance deviates from that promised in the PPA, not whether it goes to the essence of the deal. Although they are related, these two concepts are distinct. These factors do not weigh in either the common law or UCC provisions.

Excuse for nonperformance

Performance obligations may be excused if performance becomes either impossible or impracticable. Common law and the UCC treat the defense of impossibility or impracticability similarly: "Where a party's performance is made impracticable without his fault by the occurrence of an event the nonoccurrence of which was a basic assumption on which the contract was made, his duty to render that performance is discharged, unless the language or the circumstances indicate the contrary."[83] When the impracticability of performance is temporary or partial, common law treats the party as obligated to perform to the extent practicable, unless the burden of performance would be substantially increased. When the seller's capacity to deliver electricity is affected, it must allocate production and deliveries among its customers, but it may, at its option, include regular customers not then under contract, as well as its own electric requirements for further manufacture.[84]

Remedies

When a breach of a PPA contract occurs, what legal remedies are available? Money damages after the fact may not be all that an injured party seeks. Taking physical possession of an electric commodity that is in short supply may be most valuable. This may be accomplished by invoking equitable legal remedies, which are available at both common law[85] and the UCC.[86]

UCC. In other events, money damages are sought for a breach of contract. The measure of damages differs between the UCC and common law. The UCC creates distinct remedies that are available to the seller rather than the buyer; one measure of the seller's damages for nonacceptance or repudiation by the buyer is the difference between the market price of electricity *at the time and place for tender.*[87]

Common law. By contrast, common law measures the market value of the service *at the time of the breach of the PPA* in calculating the amount due the nonbreaching party. This difference in the time of measuring lost value may result in a significant difference, especially with a constantly changing electric price in a deregulated competitive retail market. Day-to-day wholesale electricity prices can change by 30% or more, depending on whether the day is hot or cold and whether it is a weekday or weekend. Therefore, there can be real differences in the amount of damages resulting from the nonbreaching contract.

UNIDROIT. Under UNIDROIT principles, the market value of the electricity not delivered is measured at the time of the breach rather than at the time the party was to perform: "Where the aggrieved party has terminated the contract and has not made a replacement transaction but there is a current price for the performance contracted for; it may recover the difference between the contract price and the price current at the time the contract is terminated as well as damages for any further harm."[88] Here it is similar to the common law time for measurement of damages.

Conclusion

The substantive law that applies to a PPA may differ dramatically depending on whether common law, the UCC, the UNIDROIT principles, or the host developing country's law is applied to adjudicate rights and disputes. These substantive differences may be outcome determinative: Which law applies may determine which party prevails and what the measure of damages is in a dispute.

Parties investing equity or debt in an SPP project dislike uncertainty. Uncertainty increases risk. Increased risk drives up the cost of financing a project and makes project completion more difficult to realize.

Particularly for lenders, who will not actively participate in the operation or management of a SPP project, but typically have more capital invested than the equity holders,[89] the legal documents must control the risk of the eventual application of unexpected law and change of law. The key legal document is the PPA. Lenders typically require that the PPA to expressly specify (1) the process for dispute resolution, (2) the language in which the resolution process will be conducted, (3) the forum or venue for dispute resolution, and (4) the substantive law to apply for dispute resolution.

International lenders typically seek an international forum, applying internationally recognized rules in an accessible language. Increasingly in the power sector, this has evolved into the application of one of the systems of laws discussed in this chapter, with an English-language preference for the tribunal. This allows lenders to be represented by their normal English-language legal counsel from international law firms and to be confident of the general system of laws to be applied to resolve disputes. Therefore, the three systems of substantive law discussed in this chapter illustrate the differences and distinctions of a typical legal playing field on which SPP projects are constructed. Similarly, international environmental principles apply, which are discussed in chapter 13.

Notes

1. The exception to this is when the Power Purchase Agreement (PPA) is determined to be a contract of adhesion and when the stronger party imposes the terms without fair bargaining. In such cases, such choice of forum clauses may be ignored.

2. United Nations Commission on International Trade Law (UNCITRAL). 1988. *Convention on the international sale of goods* (CISG), Chapter 1, Article 2(f), "This Convention does not apply to sales....[o]f electricity." The Secretariat's draft commentary summarized the reasons for the exclusion as follows: This subparagraph excludes sales of electricity from the scope of this convention on the ground that in many legal systems electricity is not considered to be goods, and in any case, international sales of electricity present unique problems that are different from those presented by the usual international sale of goods. Commentary on the Draft Convention of the International Sale of Goods U.N. Doc A/C.N. 9/116 Annex 2 (1976), art. 2, para. 10, reprinted in [1976], 7 UNCITRAL Y.B. 96, 98 U.N. Doc A/C.N. 9/Ser A/1997. Some law review articles on this subject note that electricity was excluded because many nations had not conclusively classified it as a good. Contracts for the sale of electricity are drafted in such detail as to choose the law to be applied, so that international commercial

2. legal codes are only applied by default in limited circumstances. Although the CISG does not apply to the sale of electricity, other international codes, such as the International Institute for the Unification of Private Law (UNIDROIT) principles, can be applied if the contract has failed to designate the law to apply.

3. The state of Louisiana has not adopted all of Article 2 of the Uniform Commercial Code (UCC) governing sales of goods.

4. UCC, sec. 2-105.

5. Ferrey, Steven. 2000. *The new rules: A guide to electric market regulation*, Tulsa, OK: PennWell.

6. *Soar v. National Football League Players' Ass'n*, 550 F.2d 1287, 1290 (1st Cir. 1977). "[W]hile an enforceable contract might be found in some circumstances if one or more of such questions were left unanswered, the accumulation in the instant case of so many unanswered questions is convincing evidence that there never was a consensus ad idem between the parties."

7. *Owen v. Owen*, 427 A.2d 933 (D.C. App. 1981); *Hill v. McGregor Mfg. Corp.*, 23 Mich. App. 342, 178 N.W. 2d 553 (1970).

8. American Law Institute. 1981. *Restatement (Second) of Contracts*, sec. 22. St Paul, MN: American Law Institute. According to the *Restatement*: "(1) The manifestation of mutual assent to an exchange ordinarily takes the form of an offer or proposal by one party followed by an acceptance by the other party or parties; (2) A manifestation of mutual assent may be made even though neither offer nor acceptance can be identified and even though the moment of formation cannot be determined." The *Restatement* states that even though a manifestation of intention is sought to be understood as an offer, it cannot be accepted so as to form a contract unless the terms of the contract are reasonably certain. According to *Coastland Corp. v. Third Nat'l Mtge. Co.* (611 F.2d 969 [4th Cir. 1979]), what constitutes "reasonable" certainty depends on the subject matter, purpose, and relationship of the parties and the circumstances under which the

agreement is made. A term need not be set forth with the utmost specificity; it is enough that the agreement is sufficiently clear for the court to determine the respective obligations of the parties.

9. According to the *Restatement (Second) of Contracts*, sec. 33, the terms of a contract are reasonably certain if they provide a basis for determining the existence of a breach and for giving an appropriate remedy.

10. Calamari, John D., and Joseph M. Perillo. 1999. *Contracts*, 3rd ed. St. Paul, MN: West Publishing.

11. *Pennsylvania Co. v. Wilmington Trust Co.*, 166 A. 2d 726 (1960). On this point, UCC, sec. 2-204, comment 3 states, "The more terms the parties leave open, the less likely it is that they have intended to conclude a binding agreement, but their actions may be frequently conclusive on the matter despite omissions."

12. UCC, sec. 2-305(2) states that when a price is to be fixed by the seller or by the buyer, it requires a price to be fixed in good faith. However, Subsection 4 holds that when the parties intended not to be bound unless the price was fixed or agreed and it is not fixed or agreed, there is no contract.

13. *Southwest Engineering Co. v. Martin Tractor Co.*, 473 P. 2d 18, (KS 1970) held that even the absence of a fairly important term does not necessarily make a contract fatally indefinite.

14. UCC, sec. 2-311(1).

15. UCC, sec. 2-305(1).

16. UNIDROIT, Art. 4.8(1)–(2).

17. An offer can be made irrevocable or an option created (1) by mutual consideration, (2) by statute, (3) by part performance or tender of performance under an offer for a unilateral contract, (4) under the doctrine of promissory estoppel, and (5) by a sealed instrument.

18. *Briker v. Walker*, 428 A. 2d 1129, 1130 (1981) held that "the essence of the option must be accepted according to its terms in order to generate a binding contract."

19. *Restatement (Second) of Contracts*, sec. 25.

20. According to UCC, sec. 2-205, an offer by a merchant to buy or sell goods in a signed writing, which, by its terms, gives assurances that it will be held open, is not revocable, for lack of consideration, during the time stated or it no time is stated for a reasonable time.

21. Official comment 3 to UCC, sec. 2-205: This section is intended to apply to current "firm" offers and not to long-term options and an outside limit of three months, during which such offers remain irrevocable, has been set. The three-month period during which firm offers remain irrevocable under this section need not be stated by days or date. If the offer states that it is "guaranteed" or "firm" until the occurrence of a contingency, which will occur within the three-month period, it will remain irrevocable until that event. A promise made for a longer period will operate under this section to bind the offerer only for the first three months of the period but may, of course, be renewed. If it is supported by consideration, it may continue for as long as the parties specify. This section deals only with the offer, which is not supported by consideration.

22. UNIDROIT, Art. 2.4(2)(a).

23. According to the *Restatement (Second) of Contracts*, sec. 50, acceptance of an offer is a manifestation of assent to the terms thereof made by the offeree in a manner invited or required by the offer. According to sec. 53, an offer may be accepted by the rendering of a performance only if the offer invites such an acceptance

24. The necessity of giving notice is less obvious if the offer proposes a "unilateral" contract and invites acceptance by means of performance and not a promise (*Carlill v. Carbolic Smoke Ball Co.*, 1 Q.B. 256 [1893]).

25. According to UCC, sec. 2-206(a), unless otherwise unambiguously indicated by the language or circumstances, an offer to make a contract shall be construed as inviting acceptance in any manner and by any medium reasonable under the circumstances.

26. *Adams v. Lindsell*, 106 Eng. Rep. 250 (King's Bench 1818).

27. *Restatement (Second) of Contracts*, sec. 59. In re *Pago Pago Aircrash*, 647 F.2d 704 (9th Cir. 1981); *Rorvig v. Douglas*, 123 Wn. 2d 854, 873 P.2d 492 (1994), en banc.

28. According to the *Restatement (Second) of Contracts*, sec. 39, a counter-offer is an offer made by an offeree to the offerer relating to the same matter as the original offer and proposing a substituted bargain differing from that proposed by the original offer.

29. *Gyurkey v. Babler*, 651 P. 2d 928 (1982); *Wagner v. Rainier Mfg. Co.*, 230 Or. 531, 537, 371 P. 2d 74, 77 (1962).

30. For there to be an effective, definite expression of acceptance, there must be an offer on the table. If the initial document is the seller's price quotation, subject to acceptance by the buyer, there is no offer. If the buyer follows up with a purchase order, the latter will be deemed the offerer. See *Brown Machine, Division of John Brown, Inc. v. Hercules, Inc.*, 770 S.W. 2d 416 (Mo. App. 1989); *McCarty v. Verson Allsteel Press Co.*, 89 Ill. App. 3d 498, 44 Ill. Dec. 570, 411 N.E. 2d 936 (1980).

31. UCC, sec. 2-207(1).

32. UCC, sec. 2-204(1) defines a "merchant" as a person who deals in goods of the kind or otherwise, by his occupation, holds himself out as having knowledge or skill peculiar to the practices or goods involved in the transaction, or to whom such knowledge or skill may be attributed by his employment of an agent or broker or other intermediary, who by his occupation, holds himself out as having such knowledge or skill.

33. UCC, sec. 1-204(1) states that whenever there is a requirement for any action to be taken be within a reasonable time, any time that is not manifestly unreasonable may be fixed by the agreement. Subsection 2 states that what is a reasonable time for taking any action depends on the nature, purpose, and circumstances of such action, and Subsection 3 holds that an action is taken "seasonably" when it is taken at or within the time agreed or, if no time is agreed, at or within a reasonable time.

34. UCC, sec. 2-207(2).

35. UCC, sec. 2-207(3).

36. UNIDROIT, Art. 2.11(1).

37. UNIDROIT, Art 2.11(2).

38. *Mulberry-Fairplains Water Assn. v. North Wilkesboro*, 105 N.C. App. 258, 412 S.E. 2d 910 (1992); *Ray v. Metropolitan Life Ins. Co.*, 858 F.Supp. 626 (S.D. Tex. 1994).

39. UCC, sec. 2-209(1).

40. UCC, sec. 2-209(2) provides, "A signed writing which excludes modification or rescission except by a signed writing cannot be otherwise modified or rescinded, but except as between merchants such a requirement on a form supplied by the merchant must be separately signed by the other party."

41. UNIDROIT, Art. 3.2.

42. UCC Subsection 2-209(4) is an exception to this rule, which provides that "although an attempt at modification or rescission does not satisfy the requirements of Subsection 2....it can operate as a waiver." Thus, even if the modification is not a signed writing, it may still operate as a waiver. The normal rule with respect to a waiver is that a waiver is retractable unless there is an estoppel.

43. UNIDROIT, Art. 2.18.

44. In the *Restatement (Second) of Contracts*, sec. 110, a contract for (1) a promise to answer for the debt, default, or miscarriage of another, including the promise of an executor or administrator to answer for the obligations of the decedent out of the administrator's own pocket; (2) a contract to transfer an interest in real property; (3) a contract to answer for the duty of another; (4) a contract made upon consideration of marriage; and (5) a contract that is not to be performed within one year from the making thereof must be evidenced by a writing to be enforceable.

45. As to contracts within the statute—other than UCC sec. 2-201—it has been said that the writing must contain "substantially the whole agreement and all its material terms and conditions, so that one reading it can understand from it what the agreement is" (*Mentz v. Newwitter*, 25 N.E. 1044, 1046 [N.Y. 1890]).

46. UCC, sec. 2-201(1).

47. Farnsworth, E. Allan, and William F. Young. 1995. *Contracts: Cases and materials*, 5th ed. New York: Foundation Press.

48. According to UCC, sec. 2-201(2), between merchants, if within a reasonable time a writing in confirmation of the contract and sufficient against the sender is received and the party receiving it has reason to know its contents, it satisfies the requirements of Subsection 1 against such party unless written notice of objection to its contents is given within 10 days after it is received.

49. UCC, sec. 2-201.

50. Calamari and Perillo, *Contracts*, 210.

51. Many contracts are expressed in multiple, detached documents and, if only one of them is signed, it is well to incorporate the others by reference. Absent that, "there is basic disagreement as to what constitutes a sufficient connection permitting the unsigned papers to be considered as part of the statutory memorandum" (*Marks v. Cowdin*, 123 N.E. 139 [N.Y. 1919]). Even when this is not true, the unsigned document is part of the memorandum if the documents by internal evidence refer to the same subject matter or transaction.

52. UNIDROIT, Art. 1.2.

53. "When the parties to a written contract have agreed to it as an "integration"—a complete and final embodiment of the terms of an agreement—parol evidence cannot be used to add to or vary its terms....When only part of the agreement is integrated, the same rule applies to that part, but parol evidence may be used to prove elements of the agreement not reduced to writing" (*Masterson v. Sine*, 436 P. 2d 561 [1968]).

54. Calamari and Perillo, *Contracts*, 187.

55. According to the *Restatement (Second) of Contracts*, sec. 209, (1) An integrated agreement is a writing or writings constituting the final expression of one or more terms of an agreement; (2) whether there is an integrated agreement is to be determined by the court as a question that is preliminary to determination of a question of interpretation or to application of the parol evidence rule; (3) when the parties reduce an agreement to a writing, which, in view of its completeness and specificity, reasonably appears to be a complete agreement, it is taken to be an integrated agreement unless it is established by other evidence that the writing did not constitute a final expression.

56. *Cosmopolitan Fin. Corp. v. Runnels*, 2 Haw. App. 33, 625 P. 2d 390 (1981).

57. What is "consistent" as opposed to different is a matter of different interpretation in the courts.

58. A "course of dealing" is a pattern of performance between the two parties to the contract, but it refers to how they have acted with respect to past contracts, not with respect to the contract in question. Thus, if a particular term has been used in previous contracts between the same parties and interpreted by them in a certain manner, this interpretation would be admissible to show how the term should be interpreted in the current contract (UCC, sec. 1-205 [1]).

59. A "course of performance" refers to the way the parties have conducted themselves in performing the particular contract at hand. The idea is that the parties' own actions in performing the contract supply evidence as to what they intended the contract terms to mean. (UCC sec. 2-208 [1]).

60. UCC, sec. 1-205(2) defines "usage of trade" as "any practice or method of dealing having such regularity of observance in a place, vocation or trade as to justify an expectation that it will be observed with respect to the transaction in question." The UCC requires that the trade usage have regularity of observance with respect to the transaction in question.

61. Section 203 of the *Restatement (Second) of Contracts* uses the terms "given greater weight" versus "control" in the UCC.

62. UCC, sec. 2-208(2) provides, "the express terms of the agreement and any such course of performance, as well as any course of dealing and usage of trade, shall be construed whenever reasonable as consistent with each other; but when such construction is unreasonable, express terms shall control course of performance and course of performance shall control both course of dealing and usage of trade."

63. Trade usage must be examined in its commercial setting to ascertain whether the parties wished the trade usage to take priority over the writing. See Krist, Usage of Trade and Course of Dealing, 1997 Ill. L.F. 811.

64. Calamari and Perillo, *Contracts*, 198.

65. UNIDROIT, Art. 4.3.

66. UNIDROIT, Art. 2.17.

67. UCC, sec. 2-314(1), states that unless it is excluded or modified, a warranty that the goods shall be merchantable is implied in a contract for their sale if the seller is a merchant with respect to goods of that kind. Under this section, the serving for value of food or drink to be consumed either on the premises or elsewhere

is a sale. Subsection 2 states that goods to be merchantable must at least (1) pass without objection in the trade under the contract description; and (2) in the case of fungible goods, are of fair average quality within the description; and (3) are fit for the ordinary purposes for which such goods are used; and (4) run, within the variations permitted by the agreement, of even kind, quality and quantity within each unit and among all units involved; and (5) are adequately contained, packaged, and labeled as the agreement may require; and (6) conform to the promise or affirmations of fact made on the container or label if any. Subsection 3 states that unless they are excluded or modified, other implied warranties may arise from course of dealing or usage of trade.

68. UCC, sec. 2-315 states that when the seller, at the time of contracting, has reason to know any particular purpose for which the goods are required and that the buyer is relying on the seller's skill or judgment to select or furnish suitable goods, there is, unless excluded or modified under sec.2-316, an implied warranty that the goods shall be fit for such purpose.

69. According to UCC, sec. 2-316 (2), language to exclude all implied warranties of fitness is sufficient if it states, for example, that "There are no warranties which extend beyond the description on the face hereof."

70. UNIDROIT, Art 5.6.

71. According to the *Restatement (Second) of Contracts*, sec. 251, (1) When reasonable grounds arise to believe that the obligor will command a breach by nonperformance that would of itself give the oblige a claim for damages for total breach under Section 243, the obligee may demand adequate assurance of due performance and may, if reasonable, suspend any performance for which he has not already received the agreed exchange until he receives such assurance; (2) The oblige may treat as a repudiation the obligor's failure to provide within a reasonable time such assurance of due performance as is adequate in the circumstances of the particular case.

72. The court in *Pittsburgh-Des Moines Steel Co. v. Brookhaven Manor Water Co.*, 532 F. 2d 572 (1976), held that to trigger the applicability of the statute, the expectation of due performance on the part of the other party entertained at contracting time no longer exists because of "reasonable grounds for insecurity."

73. UCC, sec. 2-609, Official comment (2): "Three measures have been adopted to meet the needs of commercial men in such situations. First, the aggrieved party is permitted to suspend his own performance and any preparation therefore, with excuse for any resulting necessary delay, until the situation has been clarified. 'Suspend performance' under this section means to hold up performance pending the outcome of the demand and includes the holding up of any preparatory action."

74. UCC, sec. 2-609, Official comment (2).

75. UCC, sec. 2-609(4): After the receipt of a justified demand, failure to provide within a reasonable time not exceeding 30 days such assurances of due performance as is adequate under the circumstances of the particular case is a repudiation of the contract.

76. UNIDROIT, Art 7.3.4.

77. Calamari and Perillo, *Contracts*, 226. In determining whether a breach of a performance obligation is material, courts will consider the following: (1) to what extent, if any, has the contract been performed at the time of breach; (2) the earlier the breach, the more likely it will be regarded as material; (3) in some jurisdictions, a willful breach is more likely to be regarded as material; (4) a degree of hardship on the breaching party; and (5) the adequacy with which the aggrieved party may be compensated by damages for partial breach. The following factors are used to determine substantial performance: (1) to what extent has the injured party obtained the benefits sought by contracting; (2) to what extent may the injured party be adequately compensated in damages; (3) to what extent has there been performance or preparation for performance; (4) how great is the hardship if the breaching party is not permitted to recover; and (5) was the breach willful?

78. *Plante v Jacobs*, 103 N.W. 2d 296 (1960), held that "There can be no recovery on the contract as distinguished from quantum meruit unless there is substantial performance. This is undoubtedly the correct rule at common law."

79. UCC, sec. 2-601.

80. UCC, sec. 2-508 provides that (1) when any tender or delivery by the seller is rejected because nonconforming and the time for performance has not yet expired, the seller may seasonably notify the buyer of his intention to cure and may then within the contract time make a conforming delivery; and (2) when the buyer rejects a nonconforming tender which the seller had reasonable grounds to believe would be acceptable with or without money allowance the seller may if he seasonably notifies the buyer have a further reasonable time to substitute a conforming tender.

81. UCC, sec.2-606 provides that (1) acceptance of goods occurs when the buyer (a) after a reasonable opportunity to inspect the goods signifies to the seller that the goods are conforming or that he will take or retain them in spite of their nonconformity; or (b) fails to make an effective rejection, but such acceptance does not occur until the buyer has had a reasonable opportunity to inspect them; or (c) does any act inconsistent with the seller's ownership; but if such act is wrongful as against the seller it is an acceptance only if ratified by him. Subsection 2 provides that acceptance of a part of any commercial unit is acceptance of that entire unit.

82. UNIDROIT, Art. 7.3.1(1)–(2).

83. UCC, sec. 2-615 (a) states a similar provision: that delay in delivery or nondelivery, in whole or in part, is not a breach of duty under a contract for sale if performance as agreed has been made impracticable by the occurrence of a contingency, the nonoccurrence of which was a basic assumption on which the contract was made, or by compliance in good faith with applicable foreign or domestic governmental regulation or order, regardless of whether it later proves to be invalid.

84. UCC, sec. 2-615 (b).

85. According to the *Restatement (Second) of Contracts*, sec. 345, the judicial remedies available for the protection of the interests stated in sec. 344 include a judgment or order (1) awarding a sum of money due under the contract or as damages, (2) requiring specific performance of a contract or enjoining its nonperformance, (3) requiring restoration of a specific thing to prevent unjust enrichment, (4) awarding a sum of money to prevent unjust enrichment, (5) declaring the rights of the parties, and (6) enforcing an arbitration award.

86. According to UCC, sec. 2-716(1), specific performance may be decreed when the goods are unique or in other proper circumstances. UCC, sec. 2-716(2), states that the decree for specific performance may include terms and conditions as to payment of the price, damages, or other relief the court deems just.

87. UCC, sec. 2-708(1) states that, subject to Subsection 2 and the provisions of this article with respect to proof of market price, the measure of damages for nonacceptance or repudiation by the buyer is the difference between the market price at the time and place for tender and the unpaid contract price together with any incidental damages provided in this article, but less expenses saved in consequence of buyer's breach.

88. 12 UNIDROIT, Art. 7.4.6(1).

89. Many projects involve 70%–80% debt financing. In such structures, the lenders have more total cash invested than the owners. Without actual operational control, lenders have only as much control as the legal documents afford them.

13 The Eye of the Needle: International Environmental Assessments and Reviews for Power Projects

As the previous chapters have illustrated, the fulcrum of any renewable or conventional energy project is financing. Much of this financing in developing countries depends on loans or loan guarantees from or supported by multilateral lending agencies. These loans and loan guarantees can be made only if the project satisfies certain lender and international criteria.

In addition to a power purchase agreement (PPA), the most challenging criterion to satisfy is the environmental review of the project. This is a significant requirement for many fossil fuel-fired power projects, and it can also challenge large renewable energy projects. Environmental assessments are now a universal prerequisite for international lending. Each project must "thread the needle" of environmental review.

Some developing nations have specific air- and water-quality requirements that power projects must satisfy. However, often these are not particularly exacting and can become even less demanding with uneven application and enforcement. International lending agencies do not require any particular air- or water-quality requirements on their own, other than that projects satisfy local permitting requirements. However, each of the international and regional lending agencies does require an independent environmental assessment and review prior to extending credit or credit guarantees. Therefore, the environmental assessment required by international agencies is the key requirement to open the tap to international lending. It is necessary to understand these elements and differences between international agencies.

Environmental assessments differ depending on which agency is involved in extending credit or support. The National Environmental Policy Act (NEPA) applies to U.S. agency involvement; each international agency has its own individual requirements. One must comply with the environmental assessment requirements of the agency from whom the small power producer (SPP) obtains credit or support. The differences are important. This chapter highlights and compares the similarities and differences in the environmental assessment requirements of different multilateral lenders as to

- Types of power projects covered

- Screening of projects for environmental assessment—different types of projects face different environmental assessment rigor

- Responsibility for environmental assessment—when independent experts and advisory panels must be used

- Initiation of the environmental assessment process—at what point in the process of project development the environmental assessment must be completed

- Project scoping—what environmental issues must be evaluated over what time period and what geographic area

- Types of environmental impacts considered—whether both positive and negative impacts must be evaluated, as well as indirect and cumulative impacts of a project

- Consideration of project alternatives—assessment of smaller scale or less environmentally damaging feasible alternatives

- Mitigation discussion or adoption—whether environmental-mitigation measures for the power project must be considered or adopted

- Timing and deadlines for of Energy Information Administration (EIA) document preparation

- Environmental assessment review and public participation—how the environmental assessment is linked to loan approval and opportunities for public comment and input

- Final decisions to proceed with energy projects—how final decisions are made

- Postapproval monitoring and auditing—whether the agency employs mechanisms to ensure environmental compliance

As a starting point, these international environmental requirements grew out of the NEPA experience in the United States. But the similarity ends there. The United States enacted NEPA in 1970. The act requires that a federal agency perform an environmental assessment for federal actions proposed, regulated, or funded by that agency. Section 102 of the act mandates that as part of the environmental assessment, the lead federal agency must prepare an environmental impact statement for all major federal actions that significantly affect the quality of the human environment. This assessment must occur prior to any agency action, decision, or commitment. This innovative requirement that agencies perform an environmental assessment before taking action has worked its way into other countries' legal systems and multinational development agencies.

Throughout the 1970s and 1980s, negative environmental consequences attributed to World Bank–funded projects prompted concern about the bank's lack of environmental review. Under pressure from nongovernmental organizations (NGOs) and other international organizations, the World Bank issued a directive in 1989 that mandated "an environmental assessment for all projects that may have a significant negative impact on the environment." Since then, other multinational development banks, such as the Asian Development Bank (ADB), the African Development Bank, and the European Development Bank, have instituted environmental assessment policies and procedures for proposed projects. Many of the current polices and procedures include the preparation of an environmental assessment report for projects that meet certain criteria.[1]

Depending on the funding source, the requirements of NEPA (United States), the ADB, or the International Bank of Reconstruction and Development (IBRD) apply. Between 1989 and 1998, these international lenders adopted environmental assessment requirements. This chapter catalogs the environmental assessment requirements instituted by NEPA, the IBRD/World Bank, the International Finance Corporation (IFC), the Multilateral Investment Guarantee Agency (MIGA), and the ADB.

Although the IBRD issued its first environmental directive in 1989, it published its operational policies and bank procedures for environmental assessment in January 1999. The IBRD's policies are not international law; rather, they are a set of guidelines and standards that the IBRD requires project sponsors to follow to receive loans for public-sector projects.

The IFC published its environmental assessment operational procedures in October 1998 and its Environmental and Social Review Procedure (ESRP) in December 1998. The IFC's environmental assessment procedures are based on the IBRD procedures, but they are not identical. Indeed, the ESRP states, "IFC's environmental and social policies, while harmonized with World Bank policies, are adapted to the private sector nature of IFC's business."[2]

The MIGA's current environmental assessment policy, which is similar to the IFC policy, became effective July 1, 1999. On February 28, 2003, the ADB issued its Operations Manual, Section 20, "Environmental Considerations in ADB Operations," which includes bank procedures and operational procedures. Therefore, for all of these international agencies, the current environmental requirements were issued between December 1998 and 2003.

Types of Projects Covered

Issue

Which international projects must undergo environmental assessment and review as a prerequisite to project financing? Project developers prefer to avoid the delay and cost of environmental reviews, and thus would prefer that a project not be covered.

NEPA. In the United States, NEPA requires an environmental assessment for all federal project and policy proposals and an environmental impact statement (EIS) for any proposals for legislation or other major federal action that significantly affect the quality of the human environment. Although these requirements appear to apply only to public projects, a federal action includes projects that are assisted, partially funded, or regulated by federal agencies; thus, many projects that are privately financed fall under the purview of NEPA.[3] There must be a major federal connection, and it must result in a significant impact on the natural environment.

IBRD. The IBRD screens every project proposed for financing to determine the appropriate extent and type of environmental assessment.[4] The IBRD funds public-sector projects only.

IFC. The IFC is a member of the World Bank Group, but, unlike the IBRD, the IFC funds private projects. The IFC requires an environmental assessment of all projects proposed for financing.[5]

MIGA. The MIGA is also a member of the World Bank Group and provides political risk guarantees to investors in projects. The MIGA requires an environmental assessment of proposed projects before providing guarantees.[6]

ADB. The ADB supports both public-sector and private-sector projects. It requires environmental assessment of all project loans, program loans, sector loans, sector development program loans, financial intermediation loans, and private-sector investment operations. The environmental assessment process covers all project components, whether they are financed by the ADB, cofinanciers, or the government.[7] The ADB treats public-sector entities somewhat differently than private-sector entities.[8]

Project Screening

Issue

Which power projects require what types of environmental assessments? Do only adverse or negative environmental impacts matter, or do significant beneficial impacts also require review? Do more sensitive environments, irreversible impacts, or novel impacts deserve special consideration?

NEPA. An agency that is considering a project uses an environmental assessment to determine whether the preparation of an environmental impact statement is required. An EIS is necessary for major federal actions that significantly affect the quality of the human environment.

The meanings of the various terms in this statement are given more precision by the Council for Environmental Quality regulations and case law. Both adverse and beneficial impacts are relevant and trigger the requirement to prepare an EIS if the impacts are significant. This differs from many of the international agency assessments, which focus only on adverse impacts.

The first step is to place the proposed project into one of three categories as to the preparation of an EIS: (1) categorically excluded, (2) categorically included, or (3) EIS in dispute. Categorically excluded projects have been predetermined to have no significant environmental impact, and therefore do not require an EIS. Categorically included projects always require an EIS. For the category of EIS in dispute, the agency performs an environmental assessment to decide whether an EIS is necessary.

IBRD, IFC, and MIGA. The IBRD, IFC, and MIGA have very similar policies for the environmental screening of proposed projects. The entity classifies the proposed project as Category A, B, C, or FI.[9] The project is classified based on the type, location, sensitivity, and scale of the project and the nature and magnitude of its potential environmental impacts.

Category A includes projects that are likely to have significant adverse environmental impacts that are sensitive, diverse, or unprecedented. Category B projects have potential adverse environmental impacts on human populations or environmentally important areas—including wetlands, forest, grasslands, and other natural habitats—that are less adverse than those of Category A projects. Category B projects have site-specific impacts, few of which are irreversible. Category C includes projects that are likely to have minimal or no adverse environmental impacts. Category FI includes subprojects for which World Bank or IFC funds are invested through a financial intermediary and for which there may be adverse environmental impacts.

Category A projects require an environmental assessment that includes a report. Category B projects require an environmental assessment that is narrower in scope than the Category A environmental assessment, and the findings and results are described in project documentation, rather than a separate environmental assessment report.

ADB. The ADB also screens projects and assigns them to Category A, B, C or FI. Similar to the IBRD, IFC, and MIGA, proposed projects are screened based on the type, location, sensitivity, scale, nature, and magnitude of potential environmental impacts. The ADB, however, also screens based on the availability of cost-effective mitigation measures.[10]

The ADB's policies provide less description of the categories than the World Bank members' policies. Category A include projects with the potential to have significant adverse environmental impacts. An environmental impact assessment is required to address significant impacts. Category B includes projects that are judged to have some adverse environmental impacts, but of lesser degree or significance than those of Category A projects. Similar to NEPA's "EIS in dispute" category, a Category B project requires an initial environmental examination to decide whether an environmental impact assessment is needed. Category C projects are those that are unlikely to have adverse environmental impacts. For Category C projects, no environmental impact assessment or initial environmental examination is needed. For these projects, the financial intermediary must apply an environmental management system, unless there are only insignificant environmental impacts. Category FI is for projects that involve a credit line through a financial intermediary or an equity investment in a financial intermediary.

Responsibility for Environmental Assessment Preparation

Issue

Who takes the lead on environmental assessments, and when must independent environmental assessors or an advisory panel be engaged for the assessment? Each agency differs.

NEPA. Under NEPA, the lead federal agency is responsible for the environmental impact statement.[11] The federal government pays the EIS preparation costs for all government-sponsored projects; for most other projects, the agency shifts responsibility and financial obligations for the environmental assessment to the private project sponsor. Federal agencies are authorized under NEPA to have regulations to recover the EIS costs of private projects, but only a minority of federal agencies have such regulations.

IBRD. After the bank performs the initial screening—that is, putting the project in Category A, B, C, or FI—the borrower is responsible for carrying out the environmental assessment.[12] For Category A projects, the borrower must retain independent environmental assessment experts who are not affiliated with the project. If a Category A project is "highly risky or contentious or [involves] serious and multidimensional environmental concerns," the bank recommends the borrower "engage an advisory panel of independent, internationally recognized environmental specialists to advise on all aspects of the project relevant to the environmental assessment."[13] Project preparation facility advances and trust funds may be available to potential borrowers that request bank assistance for financing the environmental assessment of a proposed project.[14]

IFC. The project sponsor is responsible for carrying out the environmental assessment.[15] For Category A projects, like the IBRD, IFC recommends but does not require that the project sponsor retain independent environmental assessment experts who are not affiliated with the project to prepare the environmental assessment. Also similar to the IBRD, for Category A projects that are "highly risky or contentious or that involve serious and multidimensional environmental concerns," IFC recommends but does not require that the project sponsor engage an advisory panel.[16]

MIGA. The applicant is responsible for carrying out the environmental assessment, unless the applicant is a lender or minority partner. In the case of a lender or minority partner, the applicant has to submit copies of the project sponsor's environmental assessment to the MIGA.[17] Regarding Category A projects, the MIGA has virtually identical requirements to the IBRD and IFC in terms of hiring independent experts and engaging an advisory panel.

ADB. The borrower is responsible for performing the environmental assessment in accordance with both the borrower's environmental assessment requirements and the ADB's environmental assessment requirements.[18]

Initiation of the Environmental Assessment Process

Issue

When must assessments start, and how integrated must they be with agency decision making? Generally, the applicable regulations state that environmental assessment should start as early as possible. There are, however, subtle differences.

NEPA. Agencies must integrate the NEPA process with other planning to ensure that planning and decisions reflect environmental values. There is a focus on integrating environmental assessment with planning and decision making.[19]

IBRD. The IBRD screens proposed projects at the earliest stage of the project cycle—that is, the IBRD's project cycle. This screening may occur after the bulk of project planning has been performed by the project sponsor. The IBRD Task Team, at the earliest stage of the project cycle, screens the proposed project and assigns it to one of the four categories.[20] For projects that require an environmental assessment report, the Task Team advises the borrower that before the bank can proceed to project appraisal, the environmental assessment report must be officially submitted.

IFC. The IFC's policies contain language that indicates a slightly more upstream integration of environmental assessment with other aspects of project preparation to help ensure that environmental social considerations are considered in project selection, siting, and design decisions.[21]

MIGA. The MIGA's policies do not give recommendations or requirements for an environmental assessment start time.[22]

ADB. Similar to the IBRD, the ADB's environmental assessment process is supposed to start as soon as potential projects for financing are identified. The ADB's operational procedures state that environmental assessment is ideally carried out simultaneously with the pre-feasibility and feasibility studies of the project. No deadline for completion of an environmental assessment report is given in the operational procedures; some flexibility is provided.[23]

Project Scoping

Issue

Scoping is the process of identifying issues, impacts, and alternatives for consideration during and inclusion in the environmental assessment. Project developers want the scope of adverse environmental impacts that they must review to be as narrow as possible with regard to the environmental media affected, the impacts attributed to the project, the geographic scope affected, the period that must be evaluated, and whether the project must take account of its own impacts in addition to those of other projects or emission sources in the area. There may be public input from other agencies and concerned members of the public in such determinations. The public may want the scope widened to account for a more robust assessment of impacts on an area or neighborhood.

NEPA. The lead agency determines the scope and significant issues to be analyzed in depth in the EIS.[24] Before starting the scoping process, the lead agency must publish a notice of intent to prepare an EIS. As part of the scoping process, the agency must invite the participation of federal, state, and local agencies, any affected Indian tribe, the proponent of the action, and other interested persons, including those who may not agree with the action on environmental grounds.

IBRD. The IBRD Task Team discusses the scope of the environmental assessment with the borrower, along with the procedures, schedule, and outline for any environmental assessment report that is required. According to Bank Procedure 4.01, this discussion occurs during the preparation of the Project Concept Document.[25] To prepare the Project Concept Document,

The Eye of the Needle: International Environmental Assessments and Reviews for Power Projects

the Task Team first examines the type, location, sensitivity, and scale of the proposed project, as well as the nature and magnitude of its potential impacts, in consultation with the Regional Environment Sector Unit.

Public input into the scoping process, therefore, appears limited, although the IBRD does state that the borrower should initiate consultations with project-affected groups and local NGOs about the project's environmental aspects as early as possible. For Category A projects, the borrower must consult these groups shortly after environmental screening, but before the terms of reference for the environmental assessment are finalized.

IFC. For Category A projects, the IFC has text in its operational policies similar to that of IBRD's operational procedures regarding consultation with project-affected groups shortly after environmental screening, but before the terms of reference for the environmental assessment are finalized.[26] Additionally, the IFC's ESRP states that for Category A projects, the project sponsor must consult relevant stakeholders at least twice: (1) during scoping but before the terms of reference for the environmental assessment are finalized, and (2) after a draft environmental assessment report has been prepared. The IFC has no corresponding published bank procedures, and therefore no language explicitly describing the scoping process within IFC. The IFC's ESRP states that for Category A projects, the IFC will visit the project site to determine, among other things, the issues that must be addressed in the environmental assessment.[27]

MIGA. The MIGA has language similar to the IBRD and IFC in its environmental and social review procedures regarding consultation with locally affected parties and local interest groups. The procedures state that the project sponsor should consult with local stakeholders before the terms of reference for the environmental assessment are finalized.[28] For Category A projects, the MIGA has a policy similar to the IFC, wherein the MIGA performs a site visit to determine the issues that must be addressed in the environmental assessment.[29]

ADB. The ADB employs an environmental screening categorization scheme similar to the IBRD, IFC, and MIGA, but its operations manual does not explicitly describe how scoping is to occur. In general, the borrower is responsible for doing the environmental assessment and, within the ADB, the project team is responsible for the ADB's environmental assessment process.[30] For Category A projects, one of the two required public consultations is to take place during the early stages of EIA fieldwork, although no explicit mention of scoping is present in the operations manual.[31]

Type of Impacts Considered in the Review

Issue

Do impacts include only conventional pollutants or land-use impacts on the natural environment, or do they include socioeconomic and human welfare impacts related to project development? Must indirect, cumulative, and international transboundary impacts of power projects be assessed?

NEPA. The NEPA regulations require that the EIS include impacts that are direct, indirect, or cumulative.[32] The regulations also require the EIS to have a discussion section that includes "the environmental impacts of the alternatives including the proposed action, any adverse environmental effects which cannot be avoided should the proposal be implemented, the relationship between short-term uses of man's environment and the maintenance and enhancement of long-term productivity, and any irreversible or irretrievable commitments of resources which would be involved in the proposal should it be implemented."[33]

The NEPA regulations provide eight elements that are to be included in the discussion of environmental consequences:

- Direct effects and their significance

- Indirect effects and their significance

- Possible conflicts between the proposed action and the objectives of federal, regional, state, and local (and in the case of a reservation, Indian tribe) land-use plans, policies, and controls for the area concerned

- Environmental effects of alternatives, including the proposed action

- Energy requirement and conservation potential of alternatives and mitigation measures

- Natural or depletable resource requirements and conservation potential of alternatives and mitigation measures

- Urban quality, historic and cultural resources, and the design of the built environment, including the reuse and conservation potential of alternatives and mitigation measures

- Means of mitigating adverse environmental impacts[34]

IBRD. The environmental assessment report should include a section on environmental impacts that predicts and assesses the project's likely positive and negative impacts. There also should be an identification of mitigation measures and residual negative impacts that cannot be mitigated. Additionally, there should be an exploration of opportunities for environmental enhancement.[35]

In the general discussion of the environmental assessment, the operational policies state that the natural environment (air, water, and land), human health and safety, social aspects (involuntary resettlement, indigenous peoples, and cultural property), and transboundary and global environmental aspects should be taken into account.[36] The IBRD's Operational Procedure 4.01 provides that when a project is likely to have sectoral or regional impacts, a sectoral or regional environmental assessment is required. Both regional and sectoral environmental assessments require an assessment of cumulative and indirect impacts.[37]

IFC. The IFC's operational procedures include virtually identical language to that of the IBRD operational policies regarding the types of impacts to be discussed in the environmental assessment report and the general discussion of items to be taken into account during environmental assessment.[38] It does not require an assessment of indirect or cumulative impacts, nor does it reference sectoral or regional environmental assessments.[39]

MIGA. The MIGA's operational regulations include virtually identical language to that of the IBRD operational policies regarding the general discussion of items to be taken into account during the environmental assessment.[40] However, the MIGA operational regulations do not include a section that details the contents of the environmental assessment report. The MIGA defines the environmental impact assessment as "an instrument to identify and assess the potential environmental impacts of a proposed project, evaluate alternatives, and design appropriate mitigation, management, and monitoring measures. An Environmental Action Plan is an integral part of an environmental impact assessment."[41] The agency makes no reference to cumulative or indirect impacts, and, similar to the IFC, there is no mention of sectoral or regional environmental assessments.

ADB. The ADB's requirements also do not contain a formal section describing the contents of the environmental assessment report. Regarding general requirements for environmental assessment, the ADB's operational procedures state that important considerations in undertaking the environmental assessment include identifying potential environmental impacts, including indirect and cumulative impacts, and assessing their significance.[42]

Summary. The NEPA requires consideration of direct, indirect, and cumulative impacts. The IBRD requires consideration of indirect and cumulative impacts only in sectoral and regional environmental assessments, but states that environmental assessments should take into account global and transboundary environmental aspects. The IFC states that a full, project-specific environmental impact assessment should cover direct and indirect impacts. The IFC also states that cumulative impacts may be considered as appropriate to specific projects, but discusses them only in an appendix to its procedures. The MIGA does not address indirect or cumulative impacts. The ADB specifies consideration of environmental impacts, including indirect and cumulative impacts.

Neither the IFC nor MIGA address sectoral and regional environmental assessments in their policies. Although the IFC could perform sectoral and regional environmental assessments, the MIGA is not in a position to as effectively address these broader assessments because of its later and lesser involvement in many projects.

Consideration of Project Alternatives

Issue

Alternative project development scenarios are at the core of the environmental assessment process. Are there other or better alternative locations, technologies, scales, or techniques that have lesser adverse impacts on the environment? Must the assessment quantify the environmental impact, and must it compare the impact to the environmental situation without construction of the project? Are there environmental impact thresholds that can not be crossed? Most renewable energy projects and most small SPP projects fair relatively well in such analyses.

NEPA. NEPA regulations define the section on alternatives as the heart of the EIS. The EIS needs to "rigorously explore and objectively evaluate all reasonable alternatives," including reasonable alternatives that not within the jurisdiction of the lead agency and the alternative of no action.[43] The environmental impacts of the proposal and the alternatives should be presented in comparative form to provide a clear basis for choice by the decision maker and the public.

IBRD. The IBRD requires that the "Analysis of Alternatives" section "systematically compares feasible alternatives to the proposed project site, technology, design, and operation—including the 'without project' situation—in terms of their potential environmental impacts; the feasibility of mitigating these impacts; their capital and recurrent costs; their suitability under local conditions; and their institutional, training, and monitoring requirements."[44] Additionally, environmental impacts are to be quantified to the extent possible for each alternative, and economic values are to be attached where feasible. In this section, the basis for selecting the particular project design is stated and the recommended emission levels and approaches to pollution prevention and abatement are justified.

IFC. The IFC implements language identical to that of IBRD's Annex B regarding the discussion of project alternatives.[45]

MIGA. The MIGA's operational regulations do not include a section on the contents of an environmental assessment report, and therefore do not provide an extensive description of how project alternatives are to be determined and analyzed. However, the "Environmental Screening" section of MIGA's environmental assessment policy states that "environmental assessment for a Category A project examines the project's potential negative and positive environmental impacts, compares them with those of feasible alternatives (including the 'without project' situation), and recommends any measures needed to prevent minimize, mitigate, or compensate for adverse impacts and improve environmental performance."[46]

For Category B projects, no discussion of project alternatives is presented. Rather, the project's potential negative and positive environmental impacts are to be examined and recommendations are to be made for any measure that is needed to prevent, minimize, mitigate, or compensate for adverse impacts and to improve environmental performance.

ADB. A list of important considerations for an environmental assessment includes "examining alternatives."[47]

Mitigation Discussion or Adoption of Mitigation

Issue

If a project is funded and environmental impacts result, is there an obligation to consider or implement measures to mitigate adverse environmental impacts? Do these mitigation measures become conditions of the loan covenants?

NEPA. "Appropriate mitigation measures" not already included in the proposed plan must be included in the "Alternatives" section of the environmental impact statement.[48] Mitigation measures are not required to be adopted. The lead agency's final record of decision, however, must state whether "all practicable means to avoid or minimize environmental harm from the alternative selected have been adopted, and if not, why they were not."[49] If mitigation measures are included as part of the lead agency's record of decision, the agency must condition the funding of actions on mitigation.[50]

IBRD. The environmental impact assessment report should include an Environmental Management Plan.[51] The operational policies also state that the IBRD's decision to support a project is predicated in part on the expectation that the plan will be executed effectively.[52]

IFC. The environmental assessment report should include an Environmental Action Plan.[53] The IFC predicates its decision to support a project, in part, on the expectation that the plan will be executed effectively.[54]

MIGA. The preparation of an Environmental Action Plan is not addressed separately in MIGA's policies or procedures. Unlike the IBRD and IFC, there is no separate section that describes the intent and content of such a plan.[55] In MIGA's ESRP, the section on "Revision to the Environmental Action Plan" states that for Category A projects, the Environmental Action Plan is "an essential and critical part of the environmental assessment report."[56]

ADB. The ADB's operational procedures state that designing least-cost mitigation measures and developing appropriate Environmental Management Plans are important considerations in undertaking environmental assessment.[57] Category A and environmentally sensitive Category B projects require the development of Environmental Management Plans that outline specific mitigation measures as part of the environmental assessment process. Loan agreements include specific environmental covenants, including Environmental Management Plan requirements.

Timing of EIA Document Preparation

Issue

The history of project funding and approval is replete with instances in which the environmental assessment or EIS was prepared after the project decision was made, often as an afterthought to comply with legal requirements. At what point in the funding and approval cycle must the assessment be performed? How many levels of approval are required? When in the process must it be available for public inspection?

NEPA. On a strategic level, the EIS is designed to be prepared "early enough so that it can serve practically as an important contribution to the decision-making process and will not be used to rationalize or justify decisions already made."[58] A federal agency cannot make a decision on a proposed action until at least 30 days after the publication of a notice in the *Federal Register* that an EIS has been completed and filed.[59]

IBRD. The environmental assessment report must be made available in a public place that is accessible to affected groups and local NGOs and must be officially submitted to the World Bank before the bank proceeds to project appraisal. The bank's Task Team advises the borrower of this requirement in writing.[60]

IFC. There is no specific reference to a deadline for completing or submitting an environmental assessment report in the ESRP or environmental assessment policy. Paragraph 31 of the procedure comes closest to addressing the timing of completion of the environmental assessment report: "Review of environmental and social information provided by the project sponsor normally occurs during project appraisal, although it may occur earlier in IFC's project cycle if the information is available."[61] Even the public disclosure sections do not indicate a firm deadline.

MIGA. There is no specific reference to a deadline for completing or submitting an environmental assessment report in Annex B of MIGA's operational regulations document or in MIGA's ESRP.[62] This timeline is flexible.[63] Because MIGA provides insurance rather than loans, it usually becomes involved in projects at a later point than the IBRD or IFC, and MIGA regulations provide more flexibility in the timing of environmental assessments.

ADB. The summary environmental assessment report is required to be available to the general public and circulated to the ADB's board of directors at least 120 days before the board considers the loan or, in some cases, before approval of significant changes in project scope or subprojects.[64]

Environmental Assessment Review and Public Participation

Issue

How many levels of review of the environmental assessment occur before there is a binding agency decision on the project? What is the process for public participation to critique a draft environmental assessment before it is deemed "final" and project approval and funding decisions based on it are implemented? How does it influence loan appraisal?

NEPA. Once the draft environmental impact statement is prepared and before the preparation of the final environmental impact statement, the lead agency must obtain the comments of any federal agency that has jurisdiction by law or special expertise that is relevant to any environmental impact involved or is authorized to develop and enforce environmental standards. The agency charged with preparing the EIS also must request the comments of appropriate state and local agencies, Indian tribes, and any agency that has requested it receive statements on actions of the kind proposed. The agency also must request comments from the applicant and the public.[65]

IBRD. For Category A and B projects, the Task Team and the Regional Environment Sector Unit (RESU) review the results of the environmental assessment to ensure that the environmental assessment report is consistent with the terms of reference agreed to by the borrower. For Category A projects, the appraisal mission team (that recommends specific aid packages) includes one or more environmental specialists with relevant expertise. The review gives special attention to the nature of the consultations with affected groups and local NGOs and the extent to which the views of these groups were considered. If the RESU is not satisfied with the environmental assessment, it may recommend one of three actions to regional management. It may recommend that the appraisal mission be postponed, that the mission be considered a preappraisal mission, or that certain issues be reexamined during the appraisal mission.[66] The appraisal mission normally begins after the IBRD has received and reviewed the official environmental assessment report. The RESU provides a formal clearance of the environmental aspects of the project at the decision stage. This clearance includes the treatment of the environmental aspects in the draft legal documents prepared by the Legal Department.

IFC. "Review of environmental and social information provided by the sponsor normally occurs during project appraisal, although it may occur earlier in IFC's project cycle if the information is available."[67] Category A projects include a desk review by the IFC of the environmental assessment report.[68]

MIGA. Staff undertake a detailed review of the environmental assessment and other environmental information when they receive it from the applicant.[69] For Category A projects, there is a desk review of the environmental impact assessment report.

ADB. Environment specialists in ADB regional departments review the environmental assessment reports. Quality assurance of projects and programs is performed by the project team, and formal peer review of Category A projects is performed through the ADB's environment committee.[70]

Final Decisions to Proceed with Energy Projects

Issue

Which entity makes the final decision as to whether to proceed with a proposed project?

NEPA. The agency responsible for preparing the EIS, that is, the lead agency designated for the proposed project, makes the decision as to whether to proceed with the proposed federal action or project. The agency prepares a public record of decision that states what the decision is, identifies all alternatives considered by the agency in reaching its decision, and states whether all practicable means to avoid or minimize environmental harm from the alternative selected have been adopted and, if not, why they were not.[71]

IBRD. The RESU provides formal clearance of the environmental aspects of the proposed project.[72] The IBRD's board of executive directors makes the final decision as to whether to support a proposed project.

IFC. The Environment and Social Development Department decides whether the project complies with appropriate IFC environmental and social requirements. If the department is satisfied, it sends an Environmental and Social Clearance Memorandum to the Investment Department. The department director then holds an investment review meeting, and the IFC negotiates with the project sponsor to establish the terms and conditions of IFC participation. The proposed project is then submitted to the IFC board for approval.[73]

MIGA. The MIGA decides whether to provide risk guarantees to the proposed project. After a president's report is prepared, a Risk Management Committee decides whether to recommend the proposed project for approval. Once the president's report is recommended for approval, the executive vice president, on behalf of the president, approves the president's report. This approval is subject to board concurrence.[74]

ADB. A project team (the mission) prepares an internal report and recommendation of the president that is circulated internally for a staff review meeting. After the terms and conditions of the ADB's investment are negotiated, the report and recommendation of the president is finalized and sent to the ADB's board of directors for consideration.[75]

Postapproval Monitoring and Auditing

Issue

What assurance does a lender have that the environmental representations and covenants of the energy project sponsor will be honored and effectuated in fact? Is there an ability to monitor and audit compliance with environmental conditions?

The monitoring and auditing programs differ among the entities. The IBRD, IFC, and ADB have several reporting requirements and options for monitoring the environmental aspects of project implementation. The NEPA requires only that a monitoring and enforcement program be adopted when it is applicable for mitigation. The MIGA relies primarily on warranties and representations by the applicant in loan covenants, although monitoring visits and requests for monitoring reports are permitted for Category A projects. The IBRD, IFC, MIGA, and ADB each use the World Bank's *Pollution Prevention and Abatement Handbook* as a guide to normally acceptable pollution prevention and abatement measures and emission levels. In addition, the IBRD, IFC, and ADB provide that they may accept alternative emission levels and approaches to pollution prevention and abatement, depending on national legislation and local conditions.

NEPA. There is no discussion of auditing in NEPA, and there is limited discussion of monitoring. The act does state that in its record of decision, "[a] monitoring and enforcement program shall be adopted and summarized where applicable for any mitigation."[76] Agencies may provide for monitoring to assure that their decisions are carried out and should do so in important cases.

IBRD. During the project, the borrower reports on compliance with measures agreed to with the Bank, including implementation of any Environmental Management Plan. The borrower also reports on the status of mitigation measures and the findings of monitoring programs.[77] The IBRD prepares an implementation completion report, which evaluates environmental impacts, noting whether they were anticipated in the environmental assessment report, and the effectiveness of any mitigation measures that were taken.

IFC. The IFC performs project monitoring in one or more of three ways. First, the IFC may review annual monitoring reports prepared by the project company. Second, the IFC may supervise missions carried out by the Investment Department and the Environment and Social Development Department. Third, staff of the Environment and Social Development Department may perform project site visits.[78] The IFC prepares project supervision reports annually, and these reports must include an environmental and social compliance section regarding covenants in the investment agreement.

MIGA. Postdecision monitoring is different from that of the IFC. The MIGA confirms that guarantee holders are operating in compliance with MIGA's environmental requirement through warranties and representations. For Category A projects, the agency can make requests for environmental monitoring reports or perform site visits.[79]

ADB. For Category A and environmentally sensitive Category B projects, the borrower or executing agency must submit semiannual reports on the implementation of Environmental Management Plans. This requirement is reflected in the loan agreements. Additionally, review missions from ADB regional departments conduct annual reviews of the environmental aspects of the project. The project completion report, prepared by the ADB's regional departments, includes three items related to the environmental aspects of the project. First, the report includes "a concise history of the environmental aspects of the project to completion, including an account of the performance of environmental indicators during project implementation." Second, an evaluation of the implementation of the Environmental Management Plan and environmental loan covenants is included. Third, an assessment of the performance of the executing agency is included.[80] The ADB also prepares project and program performance audit reports that include some analysis of environmental aspects of the project.[81]

Conclusion

The project sponsor must satisfy the most demanding of these environmental assessment criteria when multiple agencies fund or support a particular SPP or independent power producer project. Therefore, the way one structures a project and its financing determines which environmental clearances apply to a proposed project. Because environmental assessment and clearance must occur before a project is financed, the environmental clearances are early critical path items for the project developer.

There is an art to environmental assessment—it is more than pure environmental science. The consultants retained to perform these assessments and evaluations on behalf of the project sponsor, and how these consultants are directed and guided by the project team, are essential for the successful and timely completion of the environmental assessment or EIS. When projects are not completed, it is often because of local opposition based on environmental grounds. The environmental assessment is more than a formality for financing and support. Rather, it is a critical hurdle to successful project completion. Careful attention and coordination of environmental assessment and approvals is essential. An environmental assessment report is also referred to as "an environmental impact assessment report" or an "environmental impact statement."

Notes

1. International Finance Corporation (IFC). 1998. Environmental and Social Review Procedure.

2. *Code of Federal Regulations* (C.F.R.). 40: sec. 1508.18. Washington, DC: Government Printing Office.

3. International Bank for Reconstruction and Development (IBRD). 1999. *Operational policies* 4.01, sec. 8.

4. IFC. 1998. Operational policies 4.01, environmental assessment.

5. Multilateral Investment Guarantee Agency (MIGA). 2002. Operational Regulations, Annex B, MIGA's Environmental Assessment Policy.

6. Asian Development Bank (ADB). 2003. Operations manual, sec. 20, environmental considerations in ADB operations.

7. ADB, Operations manual, sec. 21. "Private sector entities and implementing institutions are a diverse group with varying environmental awareness and capabilities, and ADB generally adopts a flexible procedure in dealing with private sector loans and investments, to tailor environmental requirements to the investment vehicle, project and expected subprojects. Nevertheless, the substance of ADB's environmental assessment requirements for private sector investments is the same as the requirements that apply to the public sector."

8. The MIGA does not have an FI category.

9. ADB, Operations manual, sec. 6.

10. 42 U.S.C.A. sec.4332(C).

11. World Bank. 1999. *World Bank operational manual*, operational procedures 4.01, environmental assessment. Washington, DC: World Bank.

12. Ibid., sec. 4.

13. World Bank. 1999. *World Bank operational manual*, bank procedures 4.01, environmental assessment.

14. IFC, Operational policies 4.01, sec. 4.

15. Ibid., sec. 4.

16. MIGA, Operational regulations, sec. 4.

17. ADB, Operations manual, sec. 5.

18. According to 40 C.F.R. sec. 1501.2, "Agencies shall integrate the NEPA process with other planning at the earliest possible time to insure that planning and decisions reflect environmental values, to avoid delays later in the process, and to head off potential conflicts."

19. IBRD, Operational policies 4.01, sec. 2.

20. IFC, Environmental and Social Review Procedure, sec. 16. "Early start of the environmental assessment process and close integration of EA with other aspects of project preparation and early start of the EA process ensure that (a) environmental and social considerations are given adequate weight in project selection, siting, and design decisions; and (b) the review process does not delay project processing....It is the project sponsor's responsibility to prepare and submit the necessary environmental information to IFC for review during project appraisal."

21. MIGA, Operational regulations, sec. 5. "MIGA will review the findings and recommendations of the environmental assessment to determine whether they provide an adequate basis for a decision to offer a guarantee."

22. ADB, Operations manual, sec. 4. "However, environmental assessment is a process rather than a one-time report, and includes necessary environmental analyses and environmental management planning that take place throughout the project cycle."
23. 40 C.F.R. sec. 1501.7(a)(2). A detailed description of the term "scope" is provided in 40 C.F.R. sec. 1508.25, including the statement that "[s]cope consists of the range of actions, alternatives, and impacts to be considered in an environmental impact statement."
24. IBRD, Operational policies 4.01, sec. 6.
25. IFC, Operational policies 4.01, sec. 12.
26. IFC, Environmental and Social Review Procedure, sec. 32.
27. MIGA. 1999. MIGA's Environmental and Social Review Procedures. www.miga.org/creens/policies/discose/soc_rev.htm.
28. MIGA, MIGA's Environmental and Social Review Procedures, sec. 33.
29. ADB, Operations manual, sec. 5.
30. Ibid., sec. 9.
31. 40 C.F.R. sec.1508.25(c). Section 1508.8 of the NEPA regulations defines "effects" as including ecological, aesthetic, historic, cultural, economic, social, or health, whether direct, indirect, or cumulative. Effects may also include those resulting from actions that may have both beneficial and detrimental effects, even if, on balance, the agency believes that the effect will be beneficial. Direct effects are defined as including effects that are caused by the action and occur at the same time and place. Indirect effects include effects that are caused by the action and are later in time or farther removed in distance, but are still reasonably foreseeable. Indirect effects may include growth-inducing effects and other effects related to induced changes in the pattern of land use, population density, or growth rate, and related effects on air and water and other natural systems, including ecosystems.
32. 40 C.F.R. sec. 1502.16.
33. 40 C.F.R. sec. 1502.16.

34. IBRD, Operational policies 4.01, Annex A, sec. (2)(e).

35. However, no specific instructions are given in Annex B ("Contents of Environmental Assessment Report") for the discussion of transboundary and global environmental aspects.

36. World Bank. 1993. Sectoral environmental assessment. Environmental Assessment Sourcebook Update No. 4, World Bank; World Bank. 1996. Regional environmental assessment. Environmental Assessment Sourcebook Update No. 15, World Bank.

37. IFC, Operational policies 4.01, sec. 3, Annex B sec. (2)(b).

38. IFC, Environmental and Social Review Procedure, Annex C. Annex C states that a full project-specific Environmental Impact Assessment "should normally cover...(b) potential environmental and social impacts (*direct and indirect*), including opportunities for enhancement; this includes the *cumulative* impact of the proposed project and other developments which are anticipated" (emphasis added). Annex C also indicates that cumulative impacts may be considered as appropriate to specific projects. A footnote to the potential consideration of cumulative impacts states, "The assessment of cumulative impacts would take into account projects or potential developments that are realistically defined at the time the EA is undertaken, when they would directly impact on the project area."

39. MIGA, Operational regulations, Annex B, sec. 3.

40. MIGA, Operational regulations, Annex B, "Definitions."

41. ADB, Operations manual, sec. 4. In a footnote, the ADB lists the major elements typically included in an environmental assessment report. This list includes "anticipated environmental impacts and mitigation measures."

42. 40 C.F.R. sec. 1502.14.

43. IBRD, Operational policies 4.01, Annex B, sec. 2(f), "Content of an Environmental Assessment Report for a Category A Project."

44. Ibid.

45. MIGA, Operational regulations, Annex B, sec. 8(a).

46. ADB, Operations manual, sec. 4. In a footnote, the ADB's operational procedures document lists elements included in an environmental assessment report, including alternatives. There is no mention of consideration of a "without project" alternative.

47. 40 C.F.R. sec. 1502.14.

48. 40 C.F.R. sec. 1505.2.

49. 40 C.F.R. sec. 1505.3.

50. IBRD, Operational policies 4.01, Annex B, sec. 2(g), Annex C provides an Environmental Management Plan: "The EMP identifies feasible and cost-effective measures that may reduce potentially significant adverse environmental impacts to acceptable levels. The plan includes compensatory measures if mitigation measures are not feasible, cost-effective, or sufficient."

51. Ibid., Annex C, sec. 5.

52. Ibid., sec. 2(g).

53. IFC, Operational policies 4.01, Annex C, sec. 6.

54. MIGA, MIGA's Environmental and Social Review Procedures. In the "Definitions" section, the definition for environmental impact assessment states that "[a]n Environmental Action Plan is an integral part of an environmental impact assessment." The policy also includes an Environmental Action Plan in a list of different instruments that can be used to satisfy MIGA's environmental assessment requirements.

55. MIGA, MIGA's Environmental and Social Review Procedures, sec. 46. The only reference to what the Environmental Action Plan should include is in the "Definitions" section: "Environmental Action Plan: An instrument which provides details of the measures to be taken during the implementation and operation of a project to eliminate or offset adverse environmental impacts or to reduce them to acceptable levels. Included are the actions needed to implement them."

56. ADB, Operations manual, sec. 4.
57. 40 C.F.R. sec. 1502.5.
58. 40 C.F.R. sec. 1506.10(b)(2).
59. IBRD, Operational policies 4.01, sec. 9.
60. IFC, Environmental and Social Review Procedure, sec. 31.
61. MIGA, MIGA's Environmental and Social Review Procedures, sec. 8(c). The procedures state that "[w]hen an environmental impact assessment report is required, the project sponsor is required to give public notification and disclose locally, as early as possible in an appropriate manner, the environmental impact assessment report at a public place accessible to project-affected groups and local interest groups such as nongovernmental organizations." The procedures also state that the purpose of the environmental review process is to determine whether the project is in compliance with MIGA's policies, and, ideally, significant environmental issues should be addressed before submission of the president's report to the board of directors.
62. MIGA, MIGA's Environmental and Social Review Procedures, sec. 34. "The purpose of the environmental review process is to determine that either the project is in compliance with MIGA's environmental policies and consistent with the guidelines, or to suggest measure the invest must take to ensure compliance and consistency. Ideally, all significant environmental issues should be satisfactorily addressed before submission of the President's Report to the Board of Directors. However, a decision may be made, on the recommendation of MIGA Management and with Board concurrence, to issue a guarantee which is conditional on the sponsor completing necessary environmental activities or mitigation measures within a reasonable, specified time."
63. ADB, Operations manual, sec. 10; ADB. 2003. Private sector development: Strategy, policies, modalities and procedures, environmental considerations.

64. 40 C.F.R. sec. 1503.1(a)(3,4). Regarding the final EIS, federal agencies with jurisdiction by law or special expertise or agencies that are authorized to develop and enforce environmental standards must comment on an EIS within their jurisdiction, expertise, or authority. 40 C.F.R. sec. 1503.2.
65. IBRD, Operational policies 4.01, sec. 12.
66. IFC, Environmental and Social Review Procedure, sec. 31.
67. Ibid., sec. 32.
68. MIGA, MIGA's Environmental and Social Review Procedures, sec. 33.
69. ADB, Operations manual, sec. 5.
70. 40 C.F.R. sec. 1505.2.
71. IBRD, Operational policies 4.01, sec. 14.
72. IFC, Environmental and Social Review Procedure, sec. 13.
73. MIGA, MIGA's Environmental and Social Review Procedures, sec. 21–22.
74. ADB, Private sector development, 21.
75. 40 C.F.R. sec. 1505.2.
76. IBRD, Operational policies 4.01, sec. 20–21. This section has a footnote reference to "OP/BP 13.05, Project Supervision." The Task Team, in consultation with the RESU and the Legal Department, reviews the reports and determines whether the borrower's compliance is satisfactory. If it is not satisfactory, the Task Team discusses corrective actions with the borrower and follows up on the implementation of the actions.
77. IFC, Environmental and Social Review Procedure, sec. 50–51. The investment officer is responsible for ensuring that supervision reports include information on the project company's compliance with environmental and social requirements. The investment officer also makes sure that annual environmental monitoring reports are provided to the Environment and Social Development Department, which is responsible for reviewing these reports and determining

whether compliance is satisfactory. In cases of noncompliance, the Environment and Social Development Department discusses courses of action with the Investment and Legal departments and specialists in the Environment and Social Development Department. The investment officer is responsible for follow-up with the project company and the Environment and Social Development Department until the noncompliance situation is resolved.

78. MIGA, MIGA's Environmental and Social Review Procedures, sec. 48. "For all Category A projects, the guarantee holder is required to submit at MIGA's request an environmental monitoring report confirming compliance with local environmental laws and regulations, and demonstrating compliance with the Environmental Action Plan. MIGA may also carry out monitoring visits, request specific data, or carry out other measures as necessary to verify information. Frequency of site visits will depend on environmental and social complexity of the project. Evidence that a project is not in compliance are grounds for canceling coverage or denying a claim."

79. ADB, Operations manual, sec. 25–26.

80. ADB, Operations manual, sec. 26. "ADB's Operations Evaluation Department prepares project and program performance audit reports that are independent evaluations and include an analysis of the effectiveness of the EMP in achieving the intended objectives. The reports will also assess the PCR's environmental reporting for its adequacy, and focus on specific environmental issues as documented in the PCR."

Index

A

Abbreviations, xx–xxiv

Accelerated Power Development Program, 120

Acceptance of offers (electricity sale), 284:
common law, 284;
UCC standard, 284

Acid rain/acidification, 52, 63:
environmental impact, 52

Acquisition (project), 256–257

Acronyms, xx–xxiv

Acts of God, 263

Air conditioning, xi–xii, xix

Alternative renewable energy. *See* Renewable energy

Andhra Pradesh, 78–79, 125–133

Animal life, 35, 58

Arbitration (dispute resolution), 261–262

Asia (critical development), 1–34:
global warming, 1–22;
infrastructure lifetimes and choices, 22–24;
notes, 25–34

Asia dimension (global warming), 20–22

Asia financial crisis, 104, 113, 115, 118

Asian Development Bank (ADB), 232, 313–315, 317, 319–322, 324, 326–328, 330–331, 333

Assignment of rights, 260–262

Assignment or delegation (PPA), 97, 112, 133, 140, 156, 180

Avoided cost principles, 72, 191, 198–199:
updating, 198–199

Award criteria, 88, 102, 127, 135, 148, 170:
Thailand, 88;
Indonesia, 102;
India, 127, 135;
Sri Lanka, 148;
Vietnam, 170

Award data, 89, 102–103, 127, 135, 148, 170–171:
Thailand, 89;
Indonesia, 102–103;
India, 127, 135;
Sri Lanka, 148;
Vietnam, 170–171

B

Baliunas, Sallie, 13

Basic provisions (PPA), 91–92, 107, 130, 137–138, 151, 176:
parties, 91, 107, 130, 137, 151, 176;
milestones, 91, 107, 130, 137, 151, 176;
delivery of power, 91, 107, 130, 137, 151, 176;
output guarantees, 91–92, 107, 130, 137–138, 151, 176;
engineering warranties, 92, 107, 130, 138, 151, 176

Bid/bidding, 73–75, 86, 190, 193:
competitive, 73–75, 86;
Thailand, 86;
security, 190;
price, 193

Biomass energy, 6, 35, 39, 42–43, 58, 61, 89, 121:
fuel, 6, 42;
environmental impact, 39, 42;
sources, 42;
combustion process, 42

Breach of contract, 97, 111, 133, 140, 155, 179, 262–269, 273, 295–296, 307–309:
definition, 97, 111, 133, 140, 155, 179;
excuse, force majeure, and impracticability, 262–264;
change of law and regulatory out clauses, 264–266;
contract termination, 266–267;
miscellaneous provisions, 267;
remedies, 267–268, 295–296, 309;
liquidated damages, 268–269

Breach remedies, 267–268, 295–296, 309:
UCC standard, 295;
common law, 295–296;
UNIDROIT principles, 296

Byrd-Hagel Resolution, 30

C

Capacity factor, 113, 115:
limitation, 115

Capacity obligations (PPA), 95–96, 110, 133, 140, 154–155, 178–179:
duration, 95;
seasonal and hourly, 95–96

Capacity tariffs, 194

Capital costs, 233–234

Capital flows (PPA role), 275–276

Capital markets, 159–163, 216, 219:
development, 159–163, 216, 219;
flaws, 219

Carbon/carbonate, 8–9

Carbon dioxide, xi, 4, 7–15, 28, 37, 43, 48

Carbon Finance Business, 15

Carbon monoxide, 51–52

Ceylon Electricity Board (CEB), 145

Change of law, 94, 108, 113, 117, 132, 139, 153, 177, 264–266

China, xiii, 19–22, 32–33, 37

Chlorofluorocarbons (CFC), 9

Clean Air Act, 23

Climate change, 7, 11–13, 18, 62–63

Coal, xi, 7–8, 10, 20–21, 36, 68

Cogeneration, 30, 54–56, 63–64, 75, 89, 100, 115, 121

Combustion, 8–9, 19, 48–52:
combustion products, 8–9, 19

Combustion products, 8–9, 19

Commercial risk (PPA), 93, 108, 132, 139, 153, 177

Common but differentiated responsibility, 15–16

Common law, 278–281, 283–284, 286, 289, 292–293, 295–296

Index

Communist state (Vietnam), 159–169, 181:
  communist/socialist government, 159–163;
  electric sector development, 164–169, 181;
  PPAs, 162;
  SPP program, 163, 165–166

Competitive markets, 85–86, 198:
  Thailand, 85–86;
  market design, 198

Competitive solicitation, 86, 190:
  Thailand, 86

Completion ratio/failure, 90, 104, 128, 135, 149, 172:
  Thailand, 90;
  Indonesia, 104;
  India, 128, 135;
  Sri Lanka, 149;
  Vietnam, 172

Comprehensive guarantees (risk), 217

Contract enforcement (PPA), 251, 275–280:
  enforceability, 251;
  role of PPA in capital flows, 275–276;
  risks posed by single-buyer electric sectors and choice of forum, 276–278;
  whose law applies, 278–280

Contract formation (PPA), 251–257, 281–283:
  safe and reliable project operation, 251–252;
  performance incentives, 252–254;
  power sales metering, 254–255;
  grid interconnection, 255–256;
  project development milestones, 256;
  utility acquisition of project and first refusal, 256–257

Contract modification, 257–261, 286:
  key provisions, 257–258;
  fuel-adjustment clauses, 258–259;
  contract rights waiver, 260;
  assignment of rights, 260–261

Contract parties (PPA), 91, 107, 130, 137, 151, 176

Contract restatement, 298–304, 306

Contract rights, 260–261:
  assignment, 260–261;
  waiver, 260

Contract support/guarantees, 248

Contract termination, 97, 112, 133, 140, 155–156, 180, 266–267

Contract validity (PPA), 251–257:
  safe and reliable project operation, 251–252;
  performance incentives, 252–254;
  power sales metering, 254–255;
  grid interconnection, 255–256;
  project development milestones, 256;
  utility acquisition of project and first refusal, 256–257

Contractual entitlement (PPA), 69–70

Controlled solicitation, 189

Convention on the International Sale of Goods (CISG), 278

Conventional power technologies, 48–52:
  emission/emission control, 48–52;
  environmental impact, 48–52
  nitrogen oxides, 49–50;
  sulfur dioxide, 50;
  volatile organic compounds, 51;
  carbon monoxide, 51–52

Cost-based principles, 187

Course of dealing, 304

Course of performance, 305

Credit enhancements (local currency), 217

Currency risk (PPA), 93, 108, 132, 138, 152, 177

D

Debt financing, 205

Decentralization (electric generation), 54–55

Decisions to proceed (project), 330–331:
 issue, 330–331;
 NEPA, 330;
 IBRD, 331;
 IFC, 331;
 MIGA, 331;
 ADB, 331

Definitions, xx–xxiv

Delivery of power (PPA), 91, 107, 130, 137, 151, 176

Deregulation, 20

Developed countries, 3–4, 7, 14–16, 19–20, 22–23

Developing nations, xiii–xviii, 1–34, 37–46. *See individual country names*

Development (Asia), 1–34:
 global warming, 1–22;
 infrastructure lifetimes and choices, 22–24;
 notes, 25–34

Development framework (SPP), 72

Development/electrification, xiii, 2, 4–6, 22–24

Diesel engines, 62

Directory General of Electricity and Energy Development (DGEED), 99, 101

Disaggregation (PPA provisions/tariffs), xv, 76

Disintegrated/island transmission system, 68, 101, 185:
 Asia, 101, 185

Dispute adjudication, 187

Dispute resolution, xvii, 69, 97, 133, 140, 156, 180, 187, 261–262, 296–297:
 PPA, 97, 112, 133, 140, 156, 180, 187;
 adjudication, 187

Distributed energy program, xv–xvi, 55–58, 64;
 environmental impact, 64

Domestic financing, 120–125

E

East Asia financial crisis, 104, 113, 115, 118

Economic loss, 11–12

Ecosystem, xvii–xviii, 1–2, 7–14, 19, 31, 39, 46–55, 60, 62–64, 311–341

Electric Power Enterprise Permit, 102

Electric power generation, xi–xviii, 4–7, 10–11, 16, 20–21, 54–55:
 capacity, 21;
 decentralization, 54–55

Electric power generation (decentralization), 54–55:
 cogeneration, 54–55;
 environmental impact, 54–55

Electricity Act of 2003 (India), 78, 83–84, 119, 141

Electricity demand/consumption, xi–xviii, 2–6, 85, 99, 119, 145, 159–161:
 energy services, 3–6;
 Thailand, 85;
 Indonesia, 99;
 India, 119–120;
 Sri Lanka, 145;
 Vietnam, 159–161

Electricity Law (Vietnam), 164–165

Electricity of Vietnam (EVN), 82, 160–161

Electricity sales, 284:
 common law, 284;
 UCC standard, 284

Electricity/electric power, xi–xviii, 2–7, 10–11, 16, 20–24, 54–55, 85, 99, 119, 145, 159–161, 202–204, 206–208, 276–278, 284:
 demand/consumption, xi–xviii, 2–6, 85, 99, 119, 145, 159–161;
 generation, xi–xviii, 4–7, 10–11, 16, 20–21, 54–55;
 electrification, xiii, 2, 4–6, 22–24, 99, 119;

Index

investment, 202–204, 206–208, 276–278;
sales, 284

Electric-sector investment, 202–204, 206–208, 276–278:
criteria, 206–208

Electrification, xiii, 2, 4–6, 22–24, 99, 119

Emission/emission control, 8–14, 46–53:
renewable energy alternatives, 46–48;
conventional power technologies, 48–52;
acidification, 52;
fuel accidents, 53

Energy banking, 192

Energy Efficiency and Renewable Energy Network, 62

Energy Generating Authority of Thailand (EGAT), 74, 83, 85, 87–90

Energy policy, xv, 2, 16–19, 85, 99, 119–120, 145, 159, 161–163:
Thailand, 85;
Indonesia, 99;
India, 119–120;
Sri Lanka, 145;
Vietnam, 159, 161–163

Energy Sector Management Assistance Programme (ESMAP), 230

Engineering warranties (PPA), 92, 107, 130, 138, 151, 176

English law, 278

Environmental and Social Review Procedure (ESRP), 314, 321

Environmental and Sustainable Development Division (ESDD), 59

Environmental assessment initiation, 319–320:
issue, 319–320;
NEPA, 319;
IBRD, 319;
IFC, 319;
MIGA, 319;
ADB, 320

Environmental assessment preparation, 317–319:
issue, 317–319;
NEPA, 317–318;
IBRD, 318;
IFC, 318;
MIGA, 318;
ADB, 319

Environmental assessment review/public participation, 329–330:
issue, 329–330;
NEPA, 329;
IBRD, 329–330;
IFC, 330;
MIGA, 330;
ADB, 330

Environmental assessments and reviews (power projects), 311–341:
project types, 314–315;
project screening, 315–317;
responsibility for environmental assessment preparation, 317–319;
initiation of environmental assessment process, 319–320;
project scoping, 320–322;
type of impacts considered in review, 322–324;
project alternatives consideration, 325–326;
mitigation discussion or adoption, 326–327;
timing EIA document preparation, 327–328;
environmental assessment review/public participation, 329–330;
final decisions to proceed with project, 330–331;
postapproval monitoring and auditing, 331–333;
conclusion, 333–334;
notes, 334–341

Environmental impact, xvii–xviii, 1–2, 7–14, 19, 31, 39, 46–55, 60, 62–64, 311–341:
criteria, xvii;
emission/emission control, 8–14, 46–53

Environmental impact assessment (EIA) document, 327–328:
issue, 327–328;
NEPA, 328;
IBRD, 328;
IFC, 328;
MIGA, 328;
ADB, 328

Environmental impacts considered, 322–324:
issue, 322–324;
NEPA, 322–324;
IBRD, 323–324;
IFC, 323–324;
MIGA, 324;
ADB, 324

Environmental Impact Statement (EIS), 314–329

Environmental mitigation, 326–327:
issue, 326–327;
NEPA, 326;
IBRD, 327;
IFC, 327;
MIGA, 327;
ADB, 327

Equity participation, 205–206

European Union, 14

Exact performance (PPA), 293–294:
common law, 293;
UCC standard, 294;
UNIDROIT principles, 294

Exchange rates, xix

Excuse and force majeure (PPA), 94, 108–109, 113, 117, 132, 139, 153, 178, 262–264, 273, 295

Exogenous factors, 219–220

Expropriation risk, 211–212

F

Factors discouraging investment (renewable energy development), 233–237:
high capital costs, 233–234;
risk perception, 234–235;
absence of insurance instruments, 235–236;
policy options to mitigate risk, 236–237

Fairness to applicants (Indonesia), 106

Federalist system (India), 119–124

Financial catalysts, 189

Financial risk phases, 208–209

Financing, xvii–xviii, 21, 120–125, 189, 201–246:
international, xvii–xviii, 21;
domestic, 120–125;
catalysts, 189;
transition, 201–246;
risk phases, 208–209

Financing the transition, 201–246:
transitions to restructured markets, 201–202;
magnitude of required electric-sector investment, 202–204;
funding sources, 205–206;
electric-sector investment criteria, 206–208;
phases of financial risk, 208–209;
risks during design and construction, 209–210;
types of international project risk, 210–213;
shifting/bearing risk in power sector, 213–215;
risk mitigation, 215–218;
government role and reform (SPP programs), 218–220;
reform of existing legal framework, 221–222;
international credit agencies and financing functions, 223–232;
risks of renewable energy development, 233–237;
notes, 237–246

First refusal (acquisition), 256–257

Force majeure, 94, 108–109, 113, 117, 132, 139, 153, 178, 262–264, 273, 295

Index

Foreign exchange risk, 212–215, 249:
    allocation, 214–215;
    ratios, 249

Forests, 9

Forum choice, 276–278, 296–297:
    risks, 276–278

Fossil fuel, xi, xiv–xvi, xviii, 2, 6–7, 10, 16–20, 22–24, 48–52, 161:
    cost of dependence, 19–20

France, xiii

Fuel accidents, 53:
    environmental impact, 53

Fuel cells (hydrogen), 45–46:
    hydrogen, 45–46;
    hydrogen reforming, 45–56

Fuel price hedging (PPA), 96, 110, 133, 140

Fuel sources (Asia), 185

Fuel-adjustment clauses, 258–259

Fund schemes (local currency), 216–217

Funding sources, 205–206:
    debt financing, 205;
    equity participation, 205–206

G

Geothermal energy, 36–37, 44, 62:
    environmental impact, 44

Germany, xiii

Global Environment Facility (GEF), 15, 122, 229–230

Global warming science, 1–2, 7, 11–16, 18–19

Global warming, xi–xviii, 1–22:
    solutions, xiv–xviii;
    science, 1–2, 7, 11–16, 18–19;
    greenhouse gases equation, 2–7;
    greenhouse gases in atmosphere, 7–10;
    climate change, 11–13;
    institutional responses, 14–15;
    scientific uncertainty, 14–15, 18–19;
    responsibility, 15–16;
    predictions and response, 16–18;
    precautionary principle, 18–19;
    fossil fuel dependence cost, 19–20;
    Asia dimension, 20–22

Government forms (Asia), 185

Government role and reform (SPP programs), 218–220:
    successful SPP model, 218;
    market design flaws, 219;
    exogenous factors, 219–220

Greenfield renewable energy projects, 17

Greenhouse effect, 7–8

Greenhouse gases, xi–xiv, xviii, 2–10, 48:
    population, 2–4;
    development, 2, 4–6;
    electrification, 2, 4–6;
    technology, 2, 6–7;
    atmosphere, 7–10;
    greenhouse effect, 7–8

Grid interconnection, 69, 73, 94, 109, 132, 139, 153–154, 173, 178, 255–256

Grid-based power capacity, 47, 145, 157

Guarantee of payment and performance (PPA), 97, 112, 133, 140, 156, 180

H

Hansen, James, 13

Hedging (fuel price), 96, 110, 133, 140

Hexafluoride, 9

Human development index (HDI), 5

Hydrochlorofluorocarbons (HCFC), 9

Hydroelectric power, xvi, 10, 23, 37–39, 59–60, 68, 81, 119–121, 160–161, 169:
    emission/emission control, 38–39;
    environmental impact, 39, 60

Hydrofluorocarbons (HFC), 8

Hydrogen, 45–46:
    hydrogen fuel cells, 45–46:
    hydrogen reforming, 45–46

349

I

Impact types considered (environmental), 322–324:
issue, 322–324;
NEPA, 322–324;
IBRD, 323–324;
IFC, 323–324;
MIGA, 324;
ADB, 324

Impracticability (contract breach), 262–264

Indemnification, 269–270

Independent government regulator, 222

Independent power producers (IPP), 75, 82, 85, 145, 186:
Asia, 186

Indexation (tariff), 68

India renewable power program, 78–80, 83–84, 119–143:
Andhra Pradesh, 78–79, 125–133;
Tamil Nadu, 79–80, 134–140;
program overview, 119–143:
energy policy, 119–120;
Electricity Act of 2003, 78, 83–84, 119, 141;
electrification, 119;
small power producer programs, 119–125;
program coordination and integration, 120;
subsidies, 120–122;
financing, 120–125;
Andhra Pradesh, 125–133;
program design and implementation, 125–130, 134–136;
power purchase agreements, 130–133, 136–140;
Tamil Nadu, 134–140;
notes, 141–143

India, xiii, xv–xvi, 10, 20–22, 67–71, 74, 78–80, 83–86, 119–143:
Electricity Act of 2003, 78, 83–84, 119, 141;
renewable power program, 78–80, 83–84, 119–143;
program overview, 119–143

Indian Renewable Energy Development Agency (IRIDA), 120–125, 143

Indonesia, xiii, xv, 20–22, 68, 70–71, 73, 76–77, 99–118:
renewable power program, 76–77, 99–118;
program overview, 99–118

Indonesia renewable power program, 76–77, 99–118:
program overview, 99–118;
energy policy, 99;
small power provider program, 99;
power purchase agreements, 99, 107–113, 117–118;
rural electrification project loan, 99;
program design and implementation, 100–107;
notes, 114–118

Indonesian State Electricity Corporation Ltd. (PLN), 77, 99, 101

Infrastructure lifetimes and choices, xv–xviii, 22–24:
electric, xv–xviii;
lifetimes and choices, xv–xviii, 22–24;
hard infrastructure, 23

Inhofe, James, 13

Innovative measures, 199

Installed capacity (renewable technologies), 46

Institutional differences (Vietnam), 161–163

Institutional responses (global warming), 14–15:
Kyoto Protocol, 14–15

Insurance instruments, 235–236

Integrated transmission system, 68, 101, 185

Interconnection arrangements (PPA), 69, 73, 94, 109, 132, 139, 153–154, 173, 178, 189, 255–256

Intergovernmental Panel on Climate Change (IPCC), 11, 27, 29

Index

International Bank for Reconstruction and Development (IBRD), 99, 224–226, 313–316, 318–321, 323–325, 327–332:
loans, 225;
guarantees, 225–226

International Centre for Settlement of Investment Disputes (ICSID), 230

International Commission on Large Dams, 59

International credit agencies (financing functions), 223–232:
World Bank Group, 223–224;
International Bank for Reconstruction and Development, 224–226;
International Development Association, 226–227;
Multilateral Investment Guarantee Agency, 227;
International Finance Corporation, 227–228;
Prototype Carbon Fund, 228–229;
Global Environment Facility, 229–230;
Energy Sector Management Assistance Programme, 230;
International Centre for Settlement of Investment Disputes, 230;
U.S. Export-Import Bank, 231–232;
Overseas Private Investment Corporation, 232;
Asian Development Bank, 232

International Development Association (IDA), 226–227

International elements (electric sector), xvii

International Energy Agency (IEA), 4, 17, 25–26

International finance, xvii–xviii, 21, 223–232:
international credit agencies, 223–232

International Finance Corporation (IFC), 227–228, 313–316, 318–319, 321, 323–325, 327–328, 330–332

International Institute for the Unification of Private Law (UNIDROIT), 278–280, 282–283, 285–286, 288, 290–294, 296, 298

International power governance (law and principles), 275–309:
securing power contract enforcement, 275–280;
form, formation, and modification of power sale contracts, 281–288;
interpretation of deal, 288–292;
obligation to honor PPA, 292–296;
conclusion, 296–297;
notes, 297–309

International project risk types, 210–213:
sovereign, 210–211;
expropriation, 211–212;
foreign exchange, 212–213

Interpretation of deal, 288–292:
parol evidence rule, 288–291;
power warranties, 291–292

Inventory (net metering), 192

Island systems (Asia), 68, 101, 185

J–K

Kerosene-lubricant fuel, 10

Kyoto Protocol, xii–xiii, 8, 14–16, 30

L

Landfill gas, 43–44, 48:
methane, 43–44;
carbon dioxide, 43;
heating value, 43;
environmental impact, 44

Land patterns, 62–63

Law and principles (international power governance), 275–309:
securing power contract enforcement, 275–280;
form, formation, and modification of power sale contracts, 281–288;
interpretation of deal, 288–292;
obligation to honor PPA, 292–296;
conclusion, 296–297;
notes, 297–309

351

Law applications (power contract), 278–280

Legal framework reform, 221–222:
independent government regulator, 222

Legal infrastructure (SPP program design), 187–189:
dispute adjudication, 187;
cost-based principles, 187;
project enhancements, 187–188;
scale of projects, 188;
allocation of risks, 188;
interconnection, 189;
financial catalysts, 189

Legal issues, xv, xvii, 187–189, 221–222, 275–309:
legal infrastructure, 187–189;
reform, 221–222;
international power governance, 275–309;
applications, 278–280

Lenders' rights, 248, 270–271

Lessons learned (Asian SPP programs), 91, 106–107, 129–130, 136, 150, 174–175, 185–200:
Thailand, 91;
Indonesia, 106–107;
India, 129–130, 136;
Sri Lanka, 150;
Vietnam, 174–175;
government forms, 185;
fuel sources, 185;
integrated/disintegrated/island systems, 185;
SPP programs, 185–186;
power purchase agreements, 185;
tariff design issues, 186;
independent power producers, 186;
PURPA model, 186;
key issues in renewable SPP program design, 187–194;
recommended best practices and program template, 194–200;
notes, 200

Liability, 269–270:
indemnification, 269–270

Lindzen, James, 13

Liquidated damages, 268–269

Local capital markets (development), 159–163, 216, 219:
flaws, 219

Local currency, 216–217:
fund schemes, 216–217;
credit enhancements, 217

M

Macroeconomic growth, 21–22

Market design, 159–163, 201–202, 216, 219:
restructured, 201–202;
transition, 201–202;
flaws, 219

Merchantable goods, 305–306

Metering (PPA), 92, 108, 131, 138, 152, 177, 254–255

Methane, 8–9, 43–44, 48

Microturbines, 62

Milestones (PPA), 91, 107, 130, 137, 151, 176, 189, 256

Ministry of Non-Conventional Energy Sources (MNES), 120–122, 141–143

Miracle nations (Asia), 4

Mitigation (discussion or adoption), 326–327:
issue, 326–327;
NEPA, 326;
IBRD, 327;
IFC, 327;
MIGA, 327;
ADB, 327

Modernizing (Sri Lanka), 145

Monitoring/auditing (project postapproval), 331–333:
issue, 331–333;
NEPA, 332;
IBRD, 332;
IFC, 332;
MIGA, 333;
ADB, 333

Motor vehicle emissions, 10
Multilateral Investment Guarantee Agency (MIGA), 227, 313–316, 318–319, 321, 324, 326–328, 330–331, 333

N

National Environmental Policy Act (NEPA), 312–320, 322–326, 328–330, 332, 336
Natural gas, xi, 8, 36, 56, 68, 75, 89
Net metering or exchange (PPA), 93, 108, 131, 138, 152, 177, 192:
energy banking, 192;
inventory, 192
New Source Review, 23
Nitrogen oxides, 8, 44, 49–50
Nitrous oxide, 8–9, 44
Nongovernmental organizations (NGO), 313
Nonperformance excuse (PPA), 295
Number and capacity of SPPs, 88, 102, 127, 134, 148, 169–170:
Thailand, 88;
Indonesia, 102;
India, 127, 134;
Sri Lanka, 148;
Vietnam, 169–170

O

Obligation to honor PPA, 292–296:
PPA performance assurances, 292–293;
exact performance, 293–294;
excuse for nonperformance, 295;
remedies of breach, 295–296
Ocean energy, 44–45:
ocean thermal energy conversion (OTEC), 44;
ocean mechanical energy, 45;
wave energy conversion, 45
Oil fuel, xi, 8, 10, 19, 36, 68

Oil reserves, 17–18
On-grid/off-grid projects, 37
Open offers, 189
Operational obligations (PPA), 96, 111, 133, 140, 155, 179
Operational risks, 210
Option contracts, 283:
common law, 283;
UCC standard, 283;
UNIDROIT principles, 283
Organisation for Economic Co-Operation and Development (OECD), 16, 22
Organization of Petroleum Exporting Countries (OPEC), 19
Outage limitation, 113
Output guarantees (PPA), 91–92, 107, 130, 137–138, 151, 176
Overseas Private Investment Corporation (OPIC), 232
Ozone, 8

P–Q

Parol evidence rule, 288–291:
common law, 289;
UCC standard, 289–290;
UNIDROIT principles, 290–291
Partial credit guarantees, 217
Partial risk guarantees, 217
Particulate matter, 8, 64
Parties (PPA), 91, 107, 130, 137, 151, 176
Per capita energy services, 3–4, 21
Perfluorocarbons (PFC), 8
Performance assurances (PPA), 292–293:
common law, 292–293;
UCC standard, 293;
UNIDROIT principles, 293

Performance exactness (PPA), 293–294:
  common law, 293;
  UCC standard, 294;
  UNIDROIT principles, 294

Performance incentives, 99, 252–254:
  Indonesia, 99

Performance obligations (PPA), 96–97, 111–112, 133, 140, 155–156, 179–180:
  operational obligations, 96, 111, 133, 140, 155, 179;
  definition of breach, 97, 111, 133, 140, 155, 179;
  termination opportunities, 97, 112, 133, 140, 155–156, 180;
  guarantee of payment and performance, 97, 112, 133, 140, 156, 180;
  assignment or delegation, 97, 112, 133, 140, 156, 180;
  dispute resolution, 97, 112, 133, 140, 156, 180

Photovoltaic materials, 39–40, 60–61, 119–121
  photovoltaic cells, 60–61

Political risk insurance, 217

Pollutant classification, 9

Population, xii, 2–4

Postapproval monitoring/auditing (project), 331–333:
  issue, 331–333;
  NEPA, 332;
  IBRD, 332;
  IFC, 332;
  MIGA, 333;
  ADB, 333

Power authority role, 87, 101–102, 126–127, 134, 147–148, 167–169:
  Thailand, 87;
  Indonesia, 101–102;
  India, 126–127, 134;
  Sri Lanka, 147–148;
  Vietnam, 167–169

Power contract enforcement (securing), 275–280:
  role of PPA in capital flows, 275–276;
  risks posed by single-buyer electric sectors and choice of forum, 276–278;
  whose law applies, 278–280

Power purchase agreement (PPA), xv–xviii, 68–76, 78–80, 82–83, 86–97, 99–113, 117–118, 125–140, 146–157, 164–180, 185, 194, 196–198, 247–274, 281–286:
  power program design and implementation, 69–70, 86–91, 100–107, 125–130, 134–136, 146–150, 164–175, 194, 196–198, 250;
  Asia, 91–97, 99, 107–113, 117–118, 130–133, 136–140, 150–157, 175–180, 185;
  standardized, 194;
  structure, 196–198;
  key provisions, 247–274;
  formation, 281–283;
  terms added, 284–285;
  modifications, 286

Power quantity commitment (PPA), 92, 108, 131, 138, 152, 176–177

Power sale contracts (form/formation/modification), 281–288:
  formation of PPA, 281–283;
  option contracts, 283;
  acceptance of offers to sell electricity, 284;
  additional terms added to PPA, 284–285;
  modifications to PPA, 286;
  statute of frauds, 286–288

Power sale enhancements (SPP program design), 191–192:
  avoided cost principles, 191;
  renewable set-aside, 191;
  third-party sales, 192;
  net metering and energy banking, 192;
  inventory of net metering, 192

Index

Power sales metering, 69, 254–255

Power warranties, 291–292;
  UCC standard, 291–292;
  UNIDROIT principles, 292

PPA formation, 281–283:
  common law, 281;
  UCC standard, 281–282;
  UNIDROIT principles, 282–283

PPA (key provisions), 247–274:
  overview, 247–251;
  contract formation, 251–257;
  contract validity, 251–257;
  contract modification, 257–261;
  dispute resolution, 261–262;
  breach of contract, 262–269;
  liability, 269–270;
  lenders' rights, 270–271;
  notes, 271–274

PPA modifications, 286:
  common law, 286;
  UCC standard, 286;
  UNIDROIT principles, 286

PPAs (Asia), 91–97, 99, 107–113,
  130–133, 136–140, 150–157,
  175–180, 185, 194:
  Thailand, 91–97;
  Indonesia, 99, 107–113, 117–118;
  India, 130–133, 136–140;
  Sri Lanka, 150–157;
  Vietnam, 175–180;
  standardized, 194

PPAs (India), 130–133, 136–140:
  features, 130–133, 136–140;
  basic provisions, 130, 137–138;
  sale elements, 131, 138;
  risk allocation, 131–132, 138–139;
  transmission, 132, 139;
  tariff issues, 132–133, 139–140;
  performance obligations, 133, 140

PPAs (Indonesia), 99, 107–113, 117–118:
  features, 107–113;
  PPA originally agreed among stakeholders, 107–112;
  basic provisions, 107;
  sale elements, 108;
  risk allocation, 108–109;
  transmission, 109;
  tariff issues, 109–111;
  performance obligations, 111–112;
  subsequent unilateral modification of PPA, 112–113, 117–118

PPAs (Sri Lanka), 150–157:
  features, 151–157;
  basic provisions, 151;
  sale elements, 152;
  risk allocation, 152–153;
  transmission, 153–154;
  tariff issues, 154–155, 157;
  performance obligations, 155–156

PPAs (Thailand), 91–97:
  features, 91–97;
  basic provisions, 91–92;
  sale elements, 92–93;
  risk allocation, 93–94;
  transmission, 94;
  tariff issues, 94–96;
  performance obligations, 96–97

PPAs (Vietnam), 162, 166, 175–180:
  features, 176–180;
  basic provisions, 176;
  sale elements, 176–177;
  risk allocation, 177–178;
  transmission, 178;
  tariff issues, 178–179;
  performance obligations, 179–180

PPA terms added, 284–285:
  common law, 284;
  UCC standard, 285;
  UNIDROIT principles, 285

Precautionary principle, 2, 18–19, 31

Price (purchase contract), 96, 110, 133, 140, 193, 248:
  hedging, 96, 110, 133, 140;
  bidding, 193

Privatization, 74

Process transparency, 90, 104, 128, 135, 149, 172–173, 189:
  Thailand, 90;
  Indonesia, 104;
  India, 128, 135;
  Sri Lanka, 149;
  Vietnam, 172–173

Program design and implementation (India), 120, 125–130, 134–136:
coordination and integration, 120;
SPP solicitation, 126, 134;
size and resource limitations, 126, 134;
power authority role, 126–127, 134;
SPP interest and applications, 127, 134;
criteria for award, 127, 135;
award data, 127, 135;
size and type of technologies, 128, 135;
completion ratio/reasons for failure, 128, 135;
process transparency, 128, 135;
stakeholder concerns, 128–129, 135–136;
lessons, 129–130, 136

Program design and implementation (Indonesia), 100–107:
SPP solicitation, 100–101;
size and resource limitations, 101;
power authority role, 101–102;
SPP interest and applications, 102;
criteria for award, 102;
award data, 102–103;
size and type of technologies, 103;
completion ratio/reasons for failure, 104;
process transparency, 104;
stakeholder concerns, 104–105;
treatment of applicants, 105;
SPP contract after unauthorized revisions, 105;
tariff after unauthorized changes, 106;
fairness to applicants, 106;
lessons, 106–107

Program design and implementation (Sri Lanka), 146–150:
SPP solicitation, 146–147;
size and resource limitations, 147;
power authority role, 147–148;
SPP interest and applications, 148;
criteria for award, 148;
award data, 148;
size and type of technologies, 148;
completion ratio/reasons for failure, 149;
process transparency, 149;
stakeholder concerns, 149–150;
lessons, 150

Program design and implementation (Thailand), 86–91:
SPP solicitation, 86;
size and resource limitations, 86–87;
power authority role, 87;
SPP interest and applications, 88;
criteria for award, 88;
award data, 89;
size and type of technologies, 89–90;
completion ratio/reasons for failure, 90;
process transparency, 90;
stakeholder concerns, 90;
lessons, 91

Program design and implementation (Vietnam), 164–175:
Electricity Law, 164–165;
SPP solicitation of SPP, 166;
size and resource limitations, 167;
power authority and government role, 167–169;
SPP interest and applications, 169–170;
criteria for award, 170;
award data, 170–171;
size and type of technologies, 171;
completion ratio/reasons for failure, 172;
process transparency, 172–173;
stakeholder concerns, 173–174;
lessons, 174–175

Program design issues (SPP), 187–194, 198:
legal infrastructure, 187–189;
project enhancements, 187–188;
solicitation of participation and competition, 189–191;
power sale enhancements, 191–192;
tariff design, 192–194;
momentum, 198

Program designs, xv, 69–70, 72–73, 75, 77, 79–81, 86–91, 100–107, 120, 125–130, 134–136, 146–150, 164–175, 185–194, 198:
Thailand, 86–91;
Indonesia, 100–107;
India, 125–130, 134–136;
Sri Lanka, 146–150;
Vietnam, 164–175

Project alternatives consideration, 325–326:
issue, 325–326;
NEPA, 325;
IBRD, 325;
IFC, 325;
MIGA, 326;
ADB, 326

Project development milestones, 91, 107, 130, 137, 151, 176, 189, 256

Project operation (safety/reliability), 251–252

Project risk types (international), 210–213:
sovereign risk, 210–211;
expropriation risk, 211–212;
foreign exchange risk, 212–213

Project scale, 188

Project scoping (environmental assessment), 320–322:
issue, 320–322;
NEPA, 320;
IBRD, 320–321;
IFC, 321;
MIGA, 321;
ADB, 321–322

Project screening (environmental assessment/review), 315–317:
issue, 315–317;
NEPA, 315–316;
IBRD, 316;
IFC, 316;
MIGA, 316;
ADB, 317

Project types (environmental assessment/review), 314–315:
issue, 314–315;
NEPA, 314–315;
IBRD, 315;
IFC, 315;
MIGA, 315;
ADB, 315

Prototype Carbon Fund (PCF), 228–229

Public participation (environmental assessment), 329–330:
issue, 329–330;
NEPA, 329;
IBRD, 329–330;
IFC, 330;
MIGA, 330;
ADB, 330

Public Utility Regulatory Policies Act (PURPA), 69, 71, 73–74, 85, 87–88, 186:
Thailand program, 85, 87–88;
PURPA model in Asia, 186

R

Recommended best practices and program template, 194–200:
successful practices, 196–197;
PPA structure, 196–198;
competitive market design, 198;
SPP program momentum, 198;
PPA relationship incentives, 198;
avoided cost principles updating, 198–199;
proper solicitation mechanism, 199;
innovative measures, 199

Recourse to law, 249

Reform of existing legal framework, 221–222:
independent government regulator, 222

Refusal contingencies, 113, 116–117

Regional Environment Sector Unit (RESU), 329

Regulatory aspects, xv, 94, 108, 113, 117, 132, 139, 153, 177, 264–266

Regulatory out clauses, 264–266

Regulatory risk and change of law (PPA), 94, 108, 113, 117, 132, 139, 153, 177, 264–266

Relationship incentives (PPA), 198

Reliability (distributed generation), 55–58, 65

Remedies of breach, 267–268, 295–296, 309:
UCC standard, 295;
common law, 295–296;
UNIDROIT principles, 296

Renewable energy, xiv–xvi, xviii, 17–18, 20–21, 23, 35–84, 86–87, 89–90, 100–103, 191–193, 233–237:
generating equipment, xvi;
development costs, 17–18;
project financing, 21;
options/alternatives, 35–66;
resources, 35–37, 86–87, 89–90, 100–103;
technologies, 37–46;
emissions, 46–48;
power capacity, 47;
power implementation, 67–84;
set-aside, 191;
premiums, 192–193;
exemptions, 193;
development risks, 233–237

Renewable energy development, 17–18, 233–237:
costs, 17–18;
risks, 233–237;
discouraging factors, 233–237

Renewable energy options/alternatives, 35–66:
renewable source, 35–37;
developing nations, 37–46;
comparative environmental emissions, 46–53;
grid-based power capacity, 47;
environmental impact, 47–48;
decentralization of electric generation, 54–55;
distributed generation reliability, 55–58;
notes, 58–66

Renewable power implementation (overview), 67–84:
small power producer programs, 67–74;
power purchase agreement, 68–74;
Public Utility Regulatory Policies Act, 69, 71, 73;
Thailand program, 74–75;
Indonesia program, 76–77;
India program, 78–80;
Sri Lanka program, 80–81;
Vietnam program, 82–83;
notes, 83–84

Renewable resources, 35–37, 86–87, 89–90, 100–103

Renewable technologies, 37–46:
hydroelectric power, 37–39;
solar photovoltaic energy, 39–40;
wind power, 41–42;
biomass energy, 42–43;
landfill gas, 43–44;
geothermal energy, 44;
ocean energy, 44–45;
hydrogen fuel cells, 45–46;
installed capacity, 46

Responsibility (environmental assessment), 317–319:
issue, 317–319;
NEPA, 317–318;
IBRD, 318;
IFC, 318;
MIGA, 318;
ADB, 319

Restatement (contract), 298–304, 306

Restructured markets, 201–202

Rights, 247, 260–261:
developers, 247;
assignment, 260–261

Risk allocation (PPA), xvii, 69, 93–94, 108–109, 113, 117–118, 131–132, 138–139, 152–153, 177–178, 188, 215, 276–278:

sovereign risk and financial assurance, 93, 108, 113, 117–118, 131, 138, 152, 177, 215;
currency risk, 93, 108, 132, 138, 152, 177;
commercial risk, 93, 108, 132, 139, 153, 177;
regulatory risk and change of law, 94, 108, 113, 117, 132, 139, 153, 177;
excuse and force majeure, 94, 108–109, 132, 139, 153, 178

Risk mitigation, 215–218, 236–237:
developing local capital markets, 216;
local currency fund schemes, 216–217;
partial credit guarantees/local currency credit enhancements, 217;
partial risk guarantees and political risk insurance, 217;
comprehensive guarantees, 217;
tariff formulas, 218

Risk perception, 234–235

Risk shifting/bearing (power sector), 213–215:
foreign exchange risk allocation, 214–215;
sovereign risk allocation, 215

Risks (design and construction), 209–210:
start-up, 209;
operational, 210

Risks, xvii, 69, 93–94, 108–109, 113, 117–118, 131–132, 138–139, 152–153, 177–178, 188, 208–218, 233–237, 249, 264–266, 276–278:
allocation, xvii, 69, 93–94, 108–109, 113, 117–118, 131–132, 138–139, 152–153, 177–178, 188, 215, 276–278;
commercial, 93, 108, 132, 139, 153, 177;
currency, 93, 108, 132, 138, 152, 177;
regulatory, 94, 108, 113, 117, 132, 139, 153, 177, 264–266;
financial, 208–209;
design and construction, 209–210;
start-up, 209;
operational, 210;

project types, 210–213;
expropriation, 211–212;
foreign exchange, 212–215, 249;
shifting/bearing, 213–215;
mitigation, 215–218, 236–237;
tariff, 218;
development, 233–237;
perception, 234–235;
contract enforcement, 276–278;
forum choice, 276–278

Russia, 14

S

Sale elements (PPA), 92–93, 108, 131, 138, 152, 176–177, 284:
power quantity commitment, 92, 108, 131, 138, 152, 176–177;
metering, 92, 108, 131, 138, 152, 177;
net metering or exchange, 93, 108, 131, 138, 152, 177

Sale of electricity, 284:
common law, 284;
UCC standard, 284

Scientific uncertainty (global warming), 1–2, 7, 11–16, 18–19

Sea level, 11–12

Secondary greenhouse gases, 8

Securing power contract enforcement, 275–280:
role of PPA in capital flows, 275–276;
risks posed by single-buyer electric sectors and choice of forum, 276–278;
whose law applies, 278–280

Single-buyer electric sectors, 276–278

Siting and permitting (power facilities), 247

Size and resource limitations, 86–87, 101, 126, 134, 147, 167:
Thailand, 86–87;
Indonesia, 101;
India, 126, 134;
Sri Lanka, 147;
Vietnam, 167

Size and type of technologies, 89–90, 103, 128, 135, 148, 171:
    Thailand, 89–90;
    Indonesia, 103;
    India, 128, 135;
    Sri Lanka, 148;
    Vietnam, 171

Small power producer program design, 187–194, 198:
    legal infrastructure, 187–189;
    project enhancements, 187–188;
    solicitation of participation and competition, 189–191;
    power sale enhancements, 191–192;
    tariff design, 192–194;
    momentum, 198

Small power producers (SPP), xvi–xviii, 67–83, 85–91, 99, 119–125, 145, 163, 165–166, 185–194, 198, 218, 312:
    operation parameters, 69;
    program design, 72–73, 75, 77, 79–81, 86–91, 99, 119–125, 145, 163, 165–166, 187–194, 198;
    Thailand, 86–91;
    Indonesia, 99;
    India, 119–125;
    Sri Lanka, 145;
    Vietnam, 163, 165–166;
    design issues, 187–194, 198;
    model, 218

Socioeconomic system/effects, 5, 11–12, 18, 31

Solar photovoltaic energy, 23–24, 35–37, 39–40, 60–61, 119–121:
    photovoltaic materials, 39–40;
    environmental impact, 39–40;
    solar-thermal energy, 40

Solar-thermal energy, 40

Solicitation mechanism, 86, 189–190, 199:
    competitive, 86, 190;
    controlled, 189

Solicitation of participation and competition (SPP program design), 86, 189–191, 199:
    mechanism, 86, 189–190, 199;
    transparent process, 189;
    open offers, 189;
    controlled solicitation, 189;
    milestones and bid security, 190;
    competitive solicitation, 190

Solicitation of SPP, 86, 100–101, 126, 134, 146–147, 166, 189–191, 199:
    mechanism, 86, 189–190, 199;
    Thailand, 86;
    Indonesia, 100–101;
    India, 126, 134;
    Sri Lanka, 146–147;
    Vietnam, 166;
    SPP program design, 189–191

Solid waste, 8, 89, 121

Solutions (global warming), xiv–xviii

Soon, Willie, 13

Sovereign risk and financial assurance (PPA), 93, 108, 113, 118, 131, 138, 152, 177, 210–211, 215:
    enforceability, 113;
    risk allocation, 215

Sri Lanka, xvi, 67–68, 70–73, 80–81, 86, 145–157:
    renewable power program, 80–81, 145–157;
    program overview, 145–157

Sri Lanka renewable power program, 80–81, 145–157:
    program overview, 145–157;
    energy policy, 145;
    Ceylon Electricity Board, 145;
    independent power providers, 145;
    small power producers, 145;
    grid system, 145;
    program design and implementation, 146–150;
    power purchase agreements, 150–157;
    notes, 157

Stakeholder concerns, 90, 104–105, 128–129, 135–136, 149–150, 173–174, 248:
    Thailand, 90;
    Indonesia, 104–105;

Index

India, 128–129, 135–136;
Sri Lanka, 149–150;
Vietnam, 173–174

Start-up risks, 209

Statute of frauds (PPA), 286–288:
common law, 286–287;
UCC standard, 287;
general principles, 287–288;
UNIDROIT principles, 288

Subsequent modification (PPA), 112–113:
tariff, 112–113;
capacity factor, 113;
outage limitation, 113;
refusal contingencies, 113;
change of law, 113;
force majeure, 113;
sovereign risk and enforceability, 113

Subsidies (India), 120–122

Successful practices (Asia), 196–197

Suitability/financeability of SPP (Indonesia), 105

Sulfur dioxide, 8, 50

Sulfur hexafluoride, 8

Sun, 35–36

Supply resources/demand (Vietnam), 159–161

Sustainable energy program, xv–xvi, 16:
power generation technology, 16

T

Tamil Nadu, 79–80, 125, 134–140

Tariff design (SPP program design), 68–69, 186, 192–194:
Asia, 186;
renewable premiums, 192–193;
tariff floors and subsidies, 193;
price bidding, 193;
renewable exemptions, 193;
tariff incentives, 193–194;
capacity tariffs, 194;
standardized PPA, 194

Tariff floors and subsidies, 193

Tariff formulas (risk mitigation), 218

Tariff incentives, xv, 68–69, 193–194

Tariff issues (PPA), 94–96, 109–111, 132–133, 139–140, 154–155, 157, 178–179:
type of tariff, 94–95, 109, 132–133, 139, 154, 178;
capacity obligations, 95–96, 110, 133, 140, 154–155, 178–179;
fuel price hedging, 96, 110, 133, 140, 155, 179;
update mechanism, 96, 110–111, 133, 140, 155, 179;
tariff penalties for nonperformance, 96, 111, 133, 140, 155, 179

Tariff penalties for nonperformance (PPA), 96, 111, 133, 140, 155, 179

Tariffs, xv, xviii, 68–69, 71, 73, 94–96, 106, 109–113, 132–133, 139–140, 154–155, 157, 178–179, 186, 192–194, 218, 250:
incentives, xv, 68–69, 193–194;
design, 68–69, 186, 192–194;
type, 94–95, 109, 132–133, 139, 154, 178;
PPA issues, 94–96, 109–111, 132–133, 139–140, 154–155, 157, 178–179;
penalties, 96, 111, 133, 140, 155, 179;
suitability, 106;
floors and subsidies, 193;
risk mitigation formulas, 218

Tariff suitability (Indonesia), 106

Tariff type (PPA), 94–95, 109, 132–133, 139, 154, 178

Tax aspects, 87, 121:
Thailand, 87;
India, 121

Technology, 2, 6–7, 16, 37–46, 67–84

Termination opportunities (PPA), 97, 112, 133, 140, 155–156, 180, 266–267

Terminology, xx–xxiv

Thailand, xiii, xv, 22, 67–74, 85–98:
renewable power program, 74–75, 85–98;
program overview, 85–98

Thailand renewable power program, 74–75, 85–98:
program overview, 85–98;
energy policy, 85;
Electricity Generation Authority of Thailand, 85, 87–90;
generation capacity, 85;
IPP-SPP program, 85;
PURPA model, 85, 87–88;
competitive bidding, 86;
competitive solicitation process, 86;
program design and implementation, 86–91;
power purchase agreements, 91–97;
notes, 98

Third-party sales, 192

Tiger nations (Asia), 4

Tipping point (ecosystem), 12

Transition to restructured market, 201–202

Transmission (PPA), 68–69, 73, 94, 109, 132, 139, 153–154, 173, 178, 189:
interconnection arrangements, 69, 73, 94, 109, 132, 139, 153–154, 173, 178, 189;
transmission and distribution obligations, 94, 109, 132, 139, 153, 178

Transmission and distribution (T&D), 4, 23–24, 56–57, 66, 94, 109, 132, 139, 153, 178:
facilities, 4, 23–24, 56–57, 66;
networks, 4;
costs, 57–58, 66;
obligations, 94, 109, 132, 139, 153, 178

Transmission/interconnection provisions (PPA), 69, 73, 94, 109, 132, 139, 153–154, 173, 178, 189, 255–256

Treatment of applicants (Indonesia), 105

Trend (development/electrification), 4–6

Two merchants exception, 287

U

UNIDROIT. *See* International Institute for the Unification of Private Law.

Uniform Commercial Code (UCC), 278–286, 289–295, 298

United Kingdom, xiii, 7

United Nations Commission on International Trade Law, 297–298

United Nations Conference on Environment and Development, 31

United Nations Development Programme (UNDP), 5–6

United Nations Framework Convention on Climate Change (UNFCCC), 14–16, 18–19

United States, xiii, 7, 9, 14–15, 19, 21–23, 30, 75

Update mechanism (PPA), 96, 110–111, 133, 140, 155, 179

Urbanization, 21

Usage of trade, 305

U.S. Council on Environmental Quality, 14, 30

U.S. Department of Energy, 15

U.S. Export-Import Bank, 231–232

Utility acquisition (project), 256–257

Utility interconnection, 69, 73, 94, 109, 132, 139, 153–154, 173, 178, 189, 255–256

V

Vegetation, 35

Vietnam, xiii, xvi, 10, 68–69, 82–83, 159–184:
 renewable power program, 82–83 159–184;
 program overview, 159–184

Vietnam renewable power program, 82–83, 159–184:
 program overview, 159–184;
 energy policy, 159, 161–163;
 supply resources and demand, 159–161;
 institutional differences, 161–163;
 program design and implementation, 164–175;
 power purchase agreements, 175–180;
 notes, 181–184

Volatile organic compounds (VOC), 8, 51

Volcanic eruption, 58–59

W–X

Waiver of contract rights, 260

Wave energy conversion, 45

Wind power, 23, 36–37, 39, 41–42, 61, 119–121, 125:
 environmental impact, 39, 42

Wood/charcoal (biomass), 6, 89

World Bank Group, 17, 25–27, 223–224

World Energy Assessment, 3

Y–Z

Yellow dwarf star, 35–36